Physico-Chemical Principles
for Processing
of Oligomeric Blends

r Science and Engineering Monographs:
te-of-the-Art Tutorial Series

series edited by **Eli M. Pearce**, Polytechnic University, Brooklyn, New York

Associate Editors

Guennadii E. Zaikov, Russian Academy of Sciences, Moscow
Yasunori Nishijima, Kyoto University, Japan

Volume 1
Fast Polymerization Processes
Karl S. Minsker and Alexandre Al. Berlin

Volume 2
Physical Properties of Polymers: Prediction and Control
Andrey A. Askadskii

Volume 3
Multicomponent Transport in Polymer Systems for Controlled Release
Alexandre Ya. Polishchuk and Guennadii E. Zaikov

Volume 4
Physico-Chemical Principles for Processing of Oligomeric Blends
Semjon M. Mezhikovskii

In preparation

Cationic Polymerization of Olefins: Quantum Chemical Aspects
Guennadii E. Zaikov, Karl S. Minsker and Vladimir A. Babkin

This book is part of a series. The publisher will accept continuation orders which may be cancelled at any time and which provide for automatic billing and shipping of each title in the series upon publication. Please write for details.

Physico-Chemical Principles for Processing of Oligomeric Blends

Semjon M. Mezhikovskii

Institute of Chemical Physics
Russian Academy of Sciences
Moscow

Translated from the Russian
by P. Pozdeev and A. V. Vakula

Gordon and Breach Science Publishers

Australia • Canada • China • France • Germany • India •
Japan • Luxembourg • Malaysia • The Netherlands •
Russia • Singapore • Switzerland

Amsteldijk 166
1st Floor
1079 LH Amsterdam
The Netherlands

British Library Cataloguing in Publication Data

Mezhikovskii, S. M.
    Physico-chemical principles for processing of oligomeric
    blends. – (Polymer science and engineering monographs : a
    state-of-the-art tutorial series ; v. 4 – ISSN 1023-7720)
    1. Oligomers   2. Chemical engineering
    I. Title
    668.9

ISBN 90-5699-661-4

To

the memory of my teacher and friend
Alfred A. Berlin

CONTENTS

INTRODUCTION TO THE SERIES

This series will provide, in the form of single-topic volumes, state-of-the-art information in specific research areas of basic applied polymer science. Volumes may incorporate a brief history of the subject, its theoretical foundations, a thorough review of current practice and results, the relationship to allied areas, and a bibliography. Books in the series will act as authoritative references for the specialist, acquaint the non-specialist with the state of science in an allied area and the opportunity for application to his own work, and offer the student a convenient, accessible review that brings together diffuse information on a subject.

PREFACE

The once popular opinion in the 1960s and 1970s was that by the end of the twentieth century oligomers would replace high polymers as the raw material for fabrication of polymeric materials. The author was among those who shared this opinion, which has proven to be wrong. Nevertheless, in recent years oligomers have significantly reduced the role of classical thermoplastic polymers and have come to occupy a greater importance in the production of traditional materials. Oligomers have become indispensable in the production of a variety of new materials. The industrial production of oligomers has expanded markedly and continues to grow at a pace exceeding that of thermoplastics. Indeed, by the beginning of the 1990s more than 60% of polymer goods in the world market were either based on reactive oligomers or used them as a principal component. These include films, coatings, adhesives, sealants, rubber goods, bonding materials, electronics and others.

World production and consumption of the most common oligomers—oligoester maleinate/fumarates (polyester resins), oligoester epoxides (epoxy resins), oligoester (meth)acrylates (polyester acrylates), phenol-formaldehyde and similar oligomers (phenol-formaldehyde resins), isocyanates and oligoalkanediols (polyurethanes)—have already exceeded millions of tons. Production of "liquid rubbers," reactive hydrocarbon oligomers, is rapidly increasing. Although industrial production of other oligomers such as oligosiloxanes, oligoimides, oligoarylenes and oligosulfides is relatively small and varies with market demand, these oligomers are largely responsible for the progress in aerospace technology, the production of jet fuel, the development of underwater devices and materials for data recording and transmission.

Chemistry and physical chemistry of different classes of oligomers and specific features of their processing into materials were rather comprehensively described in a number of monographs and reviews. The incomplete list includes well-known monographs by J. Saunders and K. Frish (1962), A. Paken (1962), H. Boenig (1964), A. Berlin et al. (1967, 1983), M. Mogilevich et al. (1983) and M. Sobolevskii et al. (1985) that, in their time, were an important contribution to the development of oligomer science.

However, general problems of the thermodynamics of oligomeric systems and the kinetics of chemical reactions resulting in the transition of liquid (oligomer feed) to a solid body (final polymer product), which are inherent to all classes of oligomers, were not systematized. This also pertains to common and specific features of the mechanisms of formation of heterogeneous structure during the curing of various oligomers and to the correlation between the structure of initial liquid formulation and the properties of the final solid material. In the excellent book *Reactive Oligomers* by Entelis et al. (1985), in spite of the very general title, only some aspects of the synthesis of oligomers with specified functionality and methods for determining functionality type distribution and molecular weight distribution were discussed.

Moreover, there have been no serious publications devoted to oligomer blends, although during the last decade oligomer blends (oligomer–monomer, oligomer–oligomer and oligomer–polymer) have come to dominate in industrial practice. It is the oligomer blends that initiated the progress in RIM processes, interpenetrating polymer networks, hybrid binders and gradient glasses. The principle of "temporary plasticization" and other effective modification procedures in rubber and plastics technology are also associated with oligomer blends. Monographs and collections of papers dealing with this trend [e.g., *Vzaimopronikayushchie polimernye setki (Interpenetrating Polymer Networks)* by Yu. Lipatov and L. Sergeeva, 1979; *Interpenetrating Polymer Networks and Related Materials* by L. Sperling, 1981; *Reaction Injection Molding*, edited by J. Kresta, 1982, 1985; *Fizikokhimiya mnogokomponentnykh polimernykh sistem (Physical Chemistry of Multicomponent Polymer Systems)*, edited by Yu. Lipatov, 1986; and *Khimicheskoe formovanie polimerov (Reaction Molding)* by A. Malkin and V. Begishev, 1991] only partially cover the physico-chemical aspects of the problem and, moreover, mostly those aspects that are related to the final (polymeric) state of a blend.

In this monograph, we make an attempt to give a comprehensive treatment of the modern state-of-the-art in theoretical and experimental studies of oligomeric systems. The aim of this undertaking was to supply process engineers with a quantitative approach to formulation of liquid oligomer compositions and selection of the regimes for their processing to materials and articles.

From this standpoint we discuss the classification of oligomers and oligomer blends and analyze their structural organization in the liquid state. In terms of statistical thermodynamics we consider equilibrium and nonequilibrium properties of oligomeric systems. On molecular, supermolecular, topological, and colloidal levels of structural hierarchy, we trace the relationships between phase organization and the kinetics of chemical and structural transformations during the cure of oligomers. Previously reported and predicted correlations between structural parameters of cured systems and their macroscopic properties are summarized. Most sections of the book are supplemented by practical recommendations that describe the application of the discussed physico-chemical regularities to real technological practice. The last chapter deals with physico-chemical aspects of oligomer technology and materials science, including the physico-chemical analysis of the relationship between the ingredients of the formulation and the properties of the resultant material.

This book is based on research of the author and other studies of the scientific school founded by Prof. Alfred A. Berlin, conducted at the Institute of Chemical Physics of the Russian Academy of Sciences in cooperation with other research groups headed by Prof. A. Chalykh (Inst. Phys. Chem., RAS), Prof. V. Lantsov (Kuibyshev Institute of Construction Engineering), and Dr. R. Frenkel' (R&D Technological Institute of Rubber Industry). I am pleased to acknowledge their cooperation. In addition, results of studies at other research centers are also discussed—in particular, studies headed by Profs. S. Entelis, Yu. Lipatov, G. Korolev, B. Rozenberg, L. Sperling, S. Krause and T. Kwei. I have tried also to give the most comprehensive coverage of the most important results discussed at five National Conferences on Oligomer Chemistry and Physical Chemistry held in the USSR (later CIS) between 1977 and 1994 which, because of the language barrier, have been unknown to foreign scientists.

This is the reason for the prevalence of Russian editions in the lists of references. However, realizing the difficulties the reader may encounter in gaining access to the Russian publications, wherever possible I have tried to cite current monographs and reviews. The reader is referred to the original publications only where absolutely necessary.

I dedicate this book to the memory of my teacher and friend Prof. Alfred A. Berlin (1912–1978). It is my deepest belief that Alfred Berlin was one of the most outstanding chemists of the second half of this century. His contribution to oligomer science and polyconjugated systems has not been yet acknowledged by his contemporaries.

I am pleased to thank Prof. Yurii S. Lipatov, Member of the National Academy of Sciences of the Ukraine. During talks and discussions at his seminars in Kiev and Odessa some ideas that were later to become part of this book were first put into words.

It is appropriate to emphasize the fruitfulness of discussions with Profs. A. Tager, G. Korolev, V. Khozin, S. Baturin, V. Kuleznev, L. Manevich, A. Malkin, E. Oleinik, B. Rozenberg and, especially, V. Irzhak, as well as with Drs. B. Zadontsev and B. Zapadinskii.

I owe a special debt to my former students and associates: Dr. L. Abdrakhmanova, researchers Mrs. E. Vasil'chenko and L. Zhil'tsova,

Drs. A. Kotova, S. Nadzharyan, T. Repina, M. Khotimskii, Sh. Shaginyan and S. Yaroshevskii, who participated in the studies summarized in this monograph.

I am also grateful to Profs. E. Yakhnin and G. Zaikov for their special role in making this book possible.

Special thanks are due to my wife Dr. M. Tokar', who assisted in preparing the manuscript and was the first to read it. It is her caring and support that, in the final analysis, made this book a reality.

The readership of this book will include primarily research people and process engineers at scientific centers of companies involved in development and production of materials based on oligomeric and polymeric systems. It may also be of interest to scientists working in the fields of physics, chemistry, and physical chemistry of oligomers and polymers. I hope the book may be of use to university teachers and graduate students.

INTRODUCTION

Pressing, molding, extrusion, vulcanization, and some other technological operations involved in fabrication of polymer articles are performed with the purpose of forming a material, which is thereby given a shape required for a particular application.

The task of a qualified process engineer is to predict possible structural transformations that take place in various stages of the process and to control the final structure and, hence, the performance of the article by simple methods (e.g., by varying composition, temperature, pressure, concentration of active additives, and so on). Accomplishment of this task encounters many difficulties. The development of a rational technology is thus reduced to solving a simply formulated but difficult problem: to obtain a final material or an article with the best (or at least satisfying the project specification) combination of properties at a minimum consumption of raw materials, energy, and time, and to avoid environmental pollution in all stages of the production cycle. Presently, this complex problem has not been solved on a uniform basis. Neither has any unified approach been developed to solve this problem.

One of the possible approaches to optimization of polymer processing can be formulated as follows [1]. In terms of chemical cybernetics, a technological process consists in that certain information is conferred to and recorded by the material: it is this information that determines the entire combination of properties of the final material. Here, the structure of the material serves as an information carrier. In the case of polymer technologies (thermoplastic and thermosetting), the structure formation involves a sequence of physical and chemical processes, such as softening, melting, solidification, vitrification, crystallization, (co)polymerization, (co)polycondensation, cross-linking, grafting, and so on. According to the general concepts of chemical cybernetics, for a closed technological cycle $I \cdot E = $ const. where I is the amount of information contained in the initial polymer (raw material) and E is the amount of energy consumed during the processing of the polymer to the final product. This fundamental relationship bears an important consequence: the most rational technology would correspond to $E \rightarrow 0$. This implies that the process must be organized so as to provide that the final product contains maximum information with minimum energy consumed during the processing. This can be achieved only if the initial materials are polymeric systems with the highest possible degree of structural organization (high initial information level). Thus, the greater the amount of information contained in the initial polymer, the smaller the amount of information that must be added during the technological cycle and the lower the amount of energy consumed in order to obtain the required combination of properties.

For a process engineer, development of a specific process is associated with the need to solve at least two distinct problems. The first one involves appropriate choice of objects (materials) that are to be subsequently processed. These materials must possess the highest possible (under the given conditions) degree of structural organization. In particular, this feature stipulates the natural tendency to use as raw materials the structurally regular oligomers rather than the random oligomers [2], or crystallizable polymers rather than the amorphous [3]. The second problem deals with the adjustment of curing regimes that would not interfere with the order that existed in the liquid phase prior to curing (liquid-crystalline structures, cybotaxes, associates, clusters, domains, elements of short-range order, and so on; the terminology has not yet been unified) but rather allow its conservation in the solid state [4].

Solution of the first task is inseparable from the achievements in the synthesis of compounds with controlled functionality and controlled interjunction spacing [5]. It is also largely related to the presently available knowledge of physico-chemical properties of compounds used to manufacture the materials. Solution of the second problem requires a deeper insight into the effects that kinetic and thermodynamic factors have on the development of the structure of a solid [4, 6, 7]; hence, it implies the need to acquire a deeper understanding of the mechanism of "liquid–solid polymer (article)" transition. These features also explain the importance, and even the necessity, of using the principles of physico-chemical mechanics [8] in the development of rational and highly effective processes for the production of composites [9].

Next we consider these principles in greater detail, with special emphasis on their implementation in the manufacture of general-purpose polymeric materials. We also examine some applications ensuing from this analysis and perform the analysis with materials based on oligomer mixtures. We chose oligomers for analysis first, because of their importance to modern polymer industry—indeed, more than 50% of the world's polymer wares production uses oligomers—and second, because the potential conserved in the structure of oligomers has not yet been realized in the production of polymers possessing a desired combination of properties. Finally, oligomers, especially the structurally regular ones, offer a convenient model for the analysis of physico-chemical regularities that control the liquid–solid transition. We begin with a description of the basic concepts and terminology used in this monograph and give modern classification of oligomers and their mixtures. This will help process engineers feel more at ease among the variety of objects used in industrial processes.

REFERENCES

1. Frenkel', S.Ya., Polimery. Problemy, Perspectivy, Prognozy (Polymers. Problems, Prospects, Forecasts), In: *Fizika segodnya i zavtra* (Physics Today and Tomorrow), Leningrad: Nauka, 1973, pp. 176–270 (in Russian).
2. Berlin, A.A., Kefeli, T.Ya., and Korolev, G.V., *Polyefirakrilaty* (Polyester Acrylates), Moscow: Nauka, 1967 (in Russian).
3. Gaylord, N. and Mark, H., *Linear and Stereoregular Addition Polymers: Polymerization with Controlled Propagation*, New York, 1959.
4. Mezhikovskii, S.M., *Polimer–oligomernye komposity* (Polymer–Oligomer Composites), Moscow: Znanie, 1989 (in Russian).
5. Entelis, S.G., Evreinov, V.V., and Kuzaev, A.I., *Reaktsionnosposobnye oligomery* (Reactive Oligomers), Moscow: Khimiya, 1985 (in Russian).
6. Irzhak, V.I., Rozenberg, B.A., and Enikolopov, N.S., *Setchatye polimery* (Polymer Networks), Moscow: Nauka, 1979 (in Russian).
7. Berlin, A.A., Korolev, G.V., Kefeli, T.Ya., and Sivergin, Yu.M., *Akrilovye oligomery i materialy na ikh osnove* (Acrylic Oligomers and Derived Materials), Moscow: Khimiya, 1983 (in Russian).
8. Rhebinder, P.A., *Fiziko-khimicheskaya mekhanika* (Physico-Chemical Mechanics), Moscow: Nauka, 1979 (in Russian).
9. Berlin, Al.Al., Vol'fson, S.A., Oshmyan, V.G., and Enikolopov, N.S., *Printsipy sozdaniya kompozitsionnykh polimernykh materialov* (Principles of the Development of Polymer Composites), Moscow: Khimiya, 1990 (in Russian).

1 OLIGOMERS AND CHARACTERISTICS OF OLIGOMER BLENDS

1.1. OLIGOMERS: TERMINOLOGY AND CLASSIFICATION

The term "oligomer" (from Greek *oligos*, meaning several or a few, and *meros*, meaning a part or repetition) was originally introduced into scientific literature by I. Gelferich in 1930 to denote carbohydrates containing 3 to 6 monose residues. With time, the meaning of the term has considerably extended. At present, it is used to describe chemical substances occupying an intermediate position between monomers (low-molecular-mass compounds) and polymers (high-molecular-mass compounds).

The use of other terms sometimes encountered in periodic literature, such as pleinomer, synthetic resin, low-molecular-weight polymer, macromer, etc. produces terminological confusion and, what is more important, deprives oligomers of the strictly defined physical meaning as a condensed state of molecules with special properties. The special properties considered below account for the marked role that the oligomers play in many biological processes and in modern technology. For example, some cellular enzymes, polypeptide antibiotics, hormones, and other biologically active substances are in fact oligomeric compounds. The fundamental significance of oligomers as of a special condensed state of a substance is emphasized by the fact

1

that some new cosmological theories [1] give them a definite position in the evolutionary succession of molecular transformations leading eventually to the appearance of living matter.

There is a certain discrepancy between different definitions of oligomers (cf. [2, 3]). Not dwelling on the history and omitting the details of a long discussion, we will refrain to citing the commonly accepted definition given by the IUPAC Commission on the Nomenclature: "A substance composed of molecules containing a few of one or more species of atoms or groups of atoms (base units) repetitively linked to each other. The physical properties of an oligomer vary with the addition or removal of one or a few of the base units from its molecules".

The sum of base units of an oligomer is called the oligomeric block (OB). Also distinguished are the terminal groups (TG) of a molecule. Their chemical structure is different from that of the unit. The number of terminal groups in a molecule can vary (two or more), and the groups may be either like or unlike. In the simplest case, an oligomer molecule can be represented in the form

$$U \quad - \quad (X)_n \quad - \quad U_1$$
$$\text{TG} \qquad \text{OB} \qquad \text{TG}$$

where U and U_1 may be either like or unlike.

The IUPAC definition cited above is, in principle, quite accurate, although it does not specify the lower boundary between the oligomer and the monomer. Indeed, the upper boundary, which is associated with passing from compound X_n (oligomer) to compound X_{n+1} (polymer), is strictly discernible, because the next step (from X_{n+1} to X_{n+2}) is insignificant (when Δn is small, the properties of polymers are identical). In contrast, the lower boundary is not strictly specified: passing from compound X_n (oligomer) to compound X_{n-1} (monomer) exhibits no distinctive features, because both the subsequent transition to X_{n-2} and the preceding transition are associated with variations in the physical properties of substances (the properties of monomers vary even when Δn is small).

This ambiguity was eliminated by Alfred A. Berlin (1912 - 1978); creation of oligomer chemistry as of a self-contained field of science is largely associated with this name. The interpretation of monomer–oligomer–polymer hierarchy offered by A. Berlin [4] appears to be more rigorous and physically sound. The boundaries between these levels were outlined in the analysis of homologous series of organic compounds. Examining the variation of partial physical parameters $\Delta F/\Delta n$ (e.g., F may be volatility, heat capacity, density, viscosity,

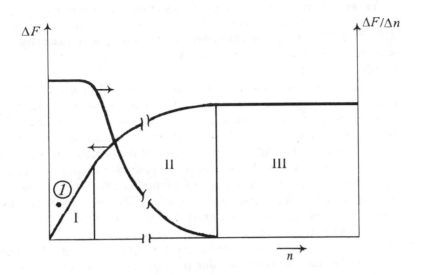

Figure 1.1. Plots of ΔF and $\Delta F/n$ versus the chain length n in a homologous series; (*1*) refers to the first member of homologous series.

melting or boiling temperature, etc.) as a function of the number of repeating units in a molecule, A. Berlin distinguished three regions in which the plots of $\Delta F/\Delta n$ versus n or ΔF versus n show markedly different patterns. For the members of homologous series, the corresponding experimentally determined physical constants were reported in [5 - 8, 12, 31, 32]. These data are generalized in Fig. 1.1.

The first region, featuring a linear dependence $\Delta F = f(n)$ (i.e., $\Delta F/\Delta n = const$), corresponds to low-molecular-mass homologs. It includes molecules with $n = 2 - 8$ (for different homologous series). The size of these molecules is smaller than the size of a segment "controlling" the property. It is noteworthy that the properties of a monomer $(n = 1)$ always fall out of the corresponding plot.

The second region shows a nonlinear pattern for the plot of partial parameter versus the number of repeating units in a molecule (i.e., $\Delta F/\Delta n \neq const$). It is this region that corresponds to the oligomers. Usually, it extends from $n = 2 - 8$ to 10^3. Oligomeric molecules may show several rotational isomers (conformers); for these, partial values of ΔF are different for molecular chains of the same length.

The third region involves high polymers. These are compounds with $n > 10^3$. They are characterized by multiple statistically averaged conformational transitions; ΔF is independent of the chain length, that is, $\Delta F / \Delta n = 0$.

The experimentally determined limiting values of n that distinguish the three regions may slightly vary even for the same homologous series (depending on what parameter F was chosen, measurement conditions, sensitivity of method, etc.). However, within the framework of the described classification, oligomers always occupy intermediate position between low-molecular-mass compounds and high polymers. Furthermore, oligomers demonstrate properties characteristic of the compounds belonging to boundary regions. Like low-molecular-mass compounds, the oligomers may exist in a molecularly dispersed form and are characterized by individual physical constants. At the same time, oligomers are characterized by rather high structural relaxation times, rather strong intermolecular interactions, low saturated vapor pressure, non-Newtonian viscosity, and other features characteristic of high polymers.

Within the framework of these definitions, many thousands of natural and synthetic compounds may be considered as oligomers. In order to systematize such a variety of compounds (it is the systematization that scientific knowledge usually involves as a first step), some essential features were specified to allow some kind of classification. The following features distinguish oligomers in technological applications.

Ability of molecules to participate (or not participate) in chemical reactions. Two types of oligomers are distinguished: (a) reactive and (b) nonreactive. This classification is conventional because the reactivity of a molecule may be different under different conditions. Indeed, a chemical bond may be involved in a reaction at certain temperatures, whereas beyond this temperature range, the bond may be inert; the reaction may take place in the presence of some catalysts, whereas other catalysts may have no effect on the reaction; a functional group may react with a certain component, whereas there is no reaction between this group and another component; etc. It is not abstract theoretical reactivity that is important for a process engineer, but the true knowledge of how oligomers will behave under definite storage, processing, and performance conditions.

Chemical nature of oligomeric block. Here, the classification is the same as in conventional polymer chemistry, and, therefore, it includes (a) carbochain, (b) heterochain, and (c) cyclic oligomers,

(d) oligomers with organoelement or inorganic chains, and others. Each class may be further subdivided to contain smaller and more specific groups of compounds, for example, oligoalkylenes, oligodienes, oligoarylenes, oligoalkylene sulfides, oligosiloxanes, etc.

Chemical nature of reactive (functional) groups. According to this feature, the following oligomers may be distinguished: (a) vinyl, (b) acrylic, (c) epoxide (glycidyl), (d) hydroxyl, (e) carboxyl, (f) isocyanate, (g) maleinate, (h) peroxide, and many others (see Table 1.1 for details).

Functionality parameters. Three functionality characteristics are used. First, the number of functional groups; second, the functionality type distribution (FTD), which also takes into account the molecular mass distribution (MMD); and third, the position of functional groups in the molecule of oligomer.

The classification according to the number of functional groups is simple: zero corresponds to nonreactive oligomers, whereas oligomers with one, two, three, four, and more functional groups are mono-, bi-, tri-, tetra-, etc. polyfunctional oligomers.

The need to classify oligomers according to FTD and MMD is associated with the fact that the main reaction leading to molecules of the required structure is usually accompanied by side reactions, which lead to imperfect molecules. For example, when a bifunctional oligomer is synthesized, oligomer molecules with the functionality f other than 2 are assumed defective. Even when present in small amounts (note that in some industrial processes the content of these by-products may be as high as 50%), these molecules may alter the properties of the oligomers and of the products of their curing [51]. Therefore, taking for granted that it is essential to conduct the synthesis of oligomers so as to obtain the products of the required structure, it is also very important (since the first requirement cannot be satisfied strictly) to know the contents of various functional groups in the resultant oligomer. The experimental methods for assessing FTD and the classification according to FTD and MMD were elaborated by Entelis [9, 50].

By analogy with the weight-average (M_w) and number-average (M_n) molecular weights, the weight-average f_w and number-average f_n functionalities are introduced

$$f_w = \sum n_i f_i^2 / \sum n_i f_i, \qquad f_n = \sum n_i f_i / \sum n_i,$$

where n_i is the number of moles of the molecules of ith type with the molecular mass M_i and functionality f_i.

Three major types of oligomers are distinguished:

(a) oligomers with strictly defined functionality, for which, in an ideal case, $f_w/f_n = 1$ and $M_w/M_n \geq 1$;

(b) oligomers, for which the relationship between M_i and f_i is linear and, at fixed values of M_n, a definite value of M_w/M_n corresponds to a definite value of f_w/f_n, that is, $f_n \propto M_n$ and $f_w/f_n \propto M_w/M_n$;

(c) polyfunctional oligomers, for which $M_w/M_n > 1$ and $f_w/f_n > 1$.

The third important functionality parameter is the position of functional groups in a molecule. For example, functional groups may belong to the main chain of oligomeric block or to the side chains; they may be located at the chain ends (telechelics); they may alternate in a regular manner, or be randomly distributed. The geometry of oligomeric molecules is essential for the molecular and supermolecular design of the final microstructure of the contrived polymer [14].

Curing pattern. Two large classes are distinguished: (a) polymerizable oligomers, the curing of which is not accompanied by the evolution of any by-products, and (b) polycondensation oligomers, which are converted to polymers by reactions that are accompanied by the liberation of low-molecular-mass compounds. Within each class, further classification may rest on the mechanism of chain growth (radical or ionic polymerization, polyaddition, migration polymerization, equilibrium or nonequilibrium polycondensation, etc.) and the mechanism of initiation or activation (peroxide, radiation-induced, catalytic, or by special curing agents, whose content in the reaction medium may be even comparable to the amount of oligomer).

Polymerizable oligomers include oligodienes, oligoester epoxides, unsaturated oligoesters, urethane-forming oligomers, and other oligomers containing multiple bonds or reactive cycles.

Polycondensation oligomers involve various phenol-formaldehyde, glyptal, and carbamide oligomers, mercaptooligosulfides, oligosiloxanediols, etc.

This classification is also conventional, because, under different conditions, one and the same functional group may react by different mechanisms (see below).

Topological structure of the cured products. For example, oligomers may lead to linear, branched, or cross-linked polymers. The other distinctive feature refers to the relaxation state of the product. For example, "liquid rubbers" are used to prepare elastomeric materials, whereas short-chain and polyfunctional oligomers lead to hard plastics.

Presently, systematics of structurally regular oligomers (oligomers with terminal or regularly alternating functional groups), which was suggested by A. Berlin in one of his last works [10], acquires permanently growing importance for process engineering. This systematics rests on a variety of relevant aspects, which include the analysis of topological structure of the cured oligomer, consideration of the chemical nature of its functional groups, assessment of the role of molecular structure of the cross-linking (curing) agents, and prediction of the possible mechanism of the cross-linking. This classification, which in fact covers most of the features discussed above, generalizes the variety of reactions that may be used to cure the oligomers; moreover, it also makes it possible to specify limitations on the application of reactive oligomers in distinct technological processes. By using this classification, a process engineer is inevitably led to make the right choice.

Table 1.1 gives an outline of the classification suggested by A.A. Berlin. The first column lists the chemical structures of the functional groups of oligomers; these structures are united into classes, which are denoted by letters from A to J (column 2). Column 3 lists the main types of oligomeric blocks associated with the particular classes of oligomers. Column 4 lists the cross-linking (curing) agents, if those are used to extend or branch the chain or form a network. For example, various amines (A) are used as curing agents for epoxy oligomers (D), although, as can be seen from the table, other chemical reactions, including those that do not use the cross-linking agents, are also possible to open the epoxide ring. Letter M denotes the nonreactive fragment lying between the functional groups of the molecule of cross-linking agent; the notation used to denote the functional groups is the same as that used with oligomer molecules (column 2). In the last two columns, the established or presumed mechanisms of oligomer curing that occur in the absence (column 5) or presence (column 6) of curing agents are described.

The table is so illustrative and simple that there is actually no need to discuss it in detail. Note, however, the important practical conclusions that follow from this table. Liquid oligomers can be converted to solid cross-linked polymers by different mechanisms by varying the nature of functional groups. By selecting appropriate methods of initiation, one and the same oligomer may be cured according to different mechanisms. Furthermore, one and the same functional group may show different functionalities. For example, the double bond F is bifunctional in chain polymerization: each scission of the double bond produces two kinetic chains and, therefore, leads to the formation of a network. However, in step polymerization (in reactions involving compounds with functional groups A and G), the double bond F is

Table 1.1. Classification of structurally regular oligomers [10]

Structure of functional group (FG)	Notation	Nature of oligomeric block (OB)	Functional groups of the cross-linking agent (CL)*	Curing mechanism without CL	Curing mechanism in the presence of CL
—OH, —H, —COOH, —NH₂	A	Oligoalkylene, oligodiene	D, E, F	–	Polyaddition
—Hal (—Br, —Cl, —F)	B	Oligoalkylene, oligodiene	C	–	Onium polymerization
—N⟨R,R (pyridine)	C	Oligoalkylene, oligodiene	B	–	Onium polymerization
epoxy —CH—CH₂, —CH—CH₂, —CH—CH₂ and the like (with O, S, NH)	D	Oligoalkylene oxide, oligoarylene oxide, oligoalkylene sulfide	A, E, or without CL	Ring-opening polymerization	Polyaddition, polymerization
—NCO (—C≡N), —NCO (—N=C=NR)	E	Oligoester, oligocarbonate, oligoamide, oligoimide	A, D, or without CL	Polymerization of isocyanates, polycyclotrimerization	Polyaddition, ionic polymerization
CH₂=CRCOO—, HOOCCH=CHCOO—, ROOCCH=CHCOO—, RC≡C—, (maleimide)	F	Oligourethane, oligosiloxane, copolymers of of different types	A, G or without CL	Multiple bond polymerization	Polyaddition, radical or ionic

			Multiple bond polymerization	Polyaddition, chain polymerization
G		A or without CL	—	
H		F		Polyaddition
I		F or without CL	—	Quasi-radical or donor-acceptor polymerization

$CH_2=CHCH_2O-$
$CH_2=CHOCO-$

$CH_2=C-$,
R

Note: *General formula of a cross-linking agent is FG—M—FG, where M is nonreactive fragment.
FG

monofunctional: only a single mobile atom migrates along the double bond and a linear polymer is formed. Another example is presented by epoxy group D, which is monofunctional in reactions with amino or carboxyl group A, whereas it is bifunctional in catalytic polymerization.

To take this feature into account, the concept of functionality in a distinct chemical reaction f_r was introduced in addition to the concept of total functionality f which refers to the number of functional groups in an ith molecule. This must be taken into account when calculating f_w and f_n.

Thus, the scheme described above makes it possible to predict the structures of products and allows a process engineer to make a purposeful choice of the starting compounds and of the procedures to be used for their conversion to polymers.

Other, less rigorous (but still important for preliminary expertise) approaches to oligomer systematization are also known. For example, reactive oligomers are also described in terms of their production volume as those produced in large amounts and in small amounts (by superficial analogy with commodity and specialty chemicals). The first are produced in amounts reaching hundreds of thousand or even millions of tons. These oligomers include unsaturated oligoesters (oligoester maleinates/fumarates and oligoester acrylates); different brands of epoxide oligomers and urethane-forming oligomers, phenol-formaldehyde and phenol-carbamide oligomers; liquid rubbers (mostly based on diene and isoprene).

The second group comprises specialty oligomers such as, for example, allylic or peroxide oligomers, vinyl oligoesters, oligosulfides, so-called "thermally stable oligomers" (when cured, these give products showing high thermal stability), and many other valuable products, whose production is not large and depends on immediate market needs.

All classifications referred to above (as virtually any other classification) are conventional and in some way deficient. Reality is much more complex than any of the schemes. This is most obviously illustrated by the modern tendency to use "mixed" oligomers containing alternating units of different chemical nature or different types of functional groups or apply oligomer-based compounds (blends) comprising oligomers of different nature and functionality (these compounds are used to prepare "hybrid binders", interpenetrating polymer networks, polymer–oligomer composites and other promising composite materials). Therefore, the definitions and the classification schemes discussed above will inevitably be improved.

1.2. OLIGOMER BLENDS

The terminology and systematics of polymer and more so of oligomer blends has not yet acquired its final organization [11 - 13, 33, 34]. Clear definitions that would allow unambiguous interpretation and would be commonly accepted have not yet been worked out. Therefore, to evade the possible misinterpretation, below we outline the physical implications of the concepts referred to in this book. However, we do not claim that the definitions given below are either comprehensive or perfect.

A blend in which one of the components is an oligomer is referred to as an oligomer blend. The second component of the blend may be represented by any low-molecular-mass, oligomeric, or high-molecular-mass compound. Therefore, the most general classification of oligomer blends involves (a) oligomer–monomer, (b) oligomer–oligomer, and (c) oligomer–polymer blends. Naturally, the classification principles that are used with individual oligomers are also valid for oligomer-based blends. However, because the number of subsystems that must be taken into account is dramatically greater, the classification principles become much more complicated.

Note also that not any of the blends, containing an oligomer and any of the second components listed above, that could be formally (sic!) classified as an oligomer blend must be considered as belonging to one of the listed classes. Indeed, formal approach would lead to that many of commercialized oligomer formulations doped with small amounts of storage inhibitors (e.g., to eliminate radical reactions, oligoester acrylates are doped with 0.1 - 1.0% of hydroquinone or other inhibitors) would have to be classified as oligomer–monomer blends. This would also be the case with oligomeric products doped with initiators (e.g., 0.1 - 3.0% of peroxides is added to oligoester acrylates), catalysts (e.g., organotin compounds are added to isocyanates) or sensibilizers (e.g., benzoquinone is added to compounds cured by light), etc. The role of these additives is very important and, sometimes, decisive for the curing process (this problem requires separate consideration); however, quite often they do not have any effect on the topological and phase organization of the liquid blends (the exceptions are specially emphasized in the text). Formally, many commercial oligomers, which are characterized by a broad distribution of oligomeric homologs, may be classified as oligomer–oligomer blends. However, the MMD and FTD of oligomers, although essential for the formation of the final structure of materials [9, 14], have almost no effect on the phase structure of the liquid oligomer mixture. This effect becomes noticeable only on micro- and submicrolevels (the terms used in [35]) of the structural organization of liquids [29, 36, 37, 47].

Therefore, when classifying oligomer blends, one has to consider primarily those factors the effect of which on the structure of initial mixture is accounted for by a significant thermodynamic contribution. These factors include the concentrations of components, chemical affinity, miscibility, etc.

Let us consider some typical examples of oligomer-based blends.

Oligomer–monomer blends. Among the vast number of oligomer–monomer blends let us mention those that are commercially important. These blends are primarily represented by mixtures of oligo(maleinate/fumarate)s with various vinyl monomers (mostly styrene); these blends are commonly referred to as polyester resins [15, 16]. Epoxide compositions (blends of oligoester epoxides with amines or other monomeric curing agents) [17, 18] and urethane composites (blends of oligoether- or oligodiene glycols with di- and triisocyanates) [19, 20] also belong to this class of oligomer blends. In different combinations these three types of blends are used in the majority (with only a few exceptions) of commercial formulations (see, e.g., [21, 22, 48, 49]).

Oligomer–oligomer blends. These include various formulations for the preparation of simultaneous interpenetrating networks (IPNs). The major requirement that these blends must satisfy is that the functional groups of each oligomer of the pair react by different reactions (this must eliminate the reaction between different chains). IPNs based on polyurethanes and polymethacrylates, polyurethanes and polyepoxides, polyacrylates and polyepoxides, polyurethanes and poly(urethane acrylate)s, and some other were described in detail [23, 24].

Another large group of oligomer–oligomer blends is represented by oligomeric compounds prepared either by mixing the available products or by synthesizing the desired oligomeric compound per se. These compounds are composed of oligomers containing chemically identical functional groups (e.g., methacrylic); however, their number in oligomer molecules is different. Furthermore, in different oligomer molecules, the nature and length of oligomer block may be either the same or different. By varying the contents of different oligomer molecules in these compounds, one can control the properties of the blends and the rigidity of the resultant network. For example, by increasing the content of the oligomer of low molecular mass one would reduce the viscosity of the resultant compound; by increasing the content of mono- or bifunctional oligomers instead of the polyfunctional oligomeric molecules one may reduce the rigidity of the resultant cross-linked polymer. A typical example of this class of oligomer oligomer blends is provided by oligoester acrylate compounds of commercial brands D-20/50 (a 1 : 1 mixture of the oligomer based on glycol and phthalic acid terminated by two methacrylic acid groups and pentaerythritol-based oligomer with eight methacrylic acid groups per oligomer molecule) and D-35 (a 35 : 65 w/w mixture of the same bifunctional oligomer and pentaerythritol-based oligomer with six methacrylic acid groups per molecule).

A popular industrial binder PN-609 based on a mixture of oligomaleinate/fumarates with an oligoether acrylate (e.g, triethylene glycol dimethacrylate) is another representative of oligomer–oligomer blends. Oligoether acrylate is used in a polyester/ether resin instead of ecologically hazardous styrene.

Mixtures of liquid reactive hydrocarbons and urethane rubbers with epoxy and unsaturated oligomers [10, 21, 38 - 44] have been permanently gaining in significance. In these blends, chain extension, branching, and cross-linking can occur simultaneously.

The examples presented demonstrate that there has been no general scheme developed yet for the application of oligomer–oligomer blends. In each particular case, a process engineer formulates an oligomer–oligomer compound almost empirically so that it fits a particular applied task.

Oligomer–polymer blends. Application of this class of oligomer-based blends suggests that at least two extreme situations are possible. In the first case, an oligomer is added to modify a polymer (let us refer to the relevant system as polymer–oligomer blend). In the second case, a linear polymer is added to a reactive oligomer* in order to make the network formed by cross-linking of polyfunctional oligomers softer and more elastic (oligomer–polymer blend). In both cases, the products obtained after curing are the polymer–polymer blends that differ in the nature of dispersion medium and the dispersed phase, morphological parameters, etc.

Polymer–oligomer blends may be further subdivided into two groups, in which the distinctive feature is the reactivity of an oligomer.

(A) When added to a high polymer, nonreactive liquid oligomers perform as plasticizers. In a resultant material, they remain chemically intact. Nonvolatility and good processibility are their main advan-

---

* Adding a polymer to a nonreactive oligomer is meaningless and, therefore, has not been implemented.

tages over common monomeric plasticizers. Oligomeric plasticizers are widely used in rubber and film technology [25 - 28].

(B) When a polymer is added to a reactive oligomer, the so-called "temporary plasticization" takes place [4]. During the initial processing stages of a polymer–oligomer blend (mixing of components, rolling, extrusion, etc.), the reactive oligomer performs as a common plasticizer (i.e., its effect is similar to that of nonreactive oligomer) and lowers the softening and flow temperatures, reduces the viscosity of the blend, etc. However, during further processing (molding) after triggering the chemical reactions, a reactive oligomer (in contrast to nonreactive plasticizers) undergoes chemical transformations (either a network is formed or graft polymerization to a linear matrix takes place) that result in the modification of polymer substrate. Thus, temporary plasticization suggests that both physical and chemical modification of linear polymers can be combined in a single technological process, and this possibility makes polymer processing more flexible.

The principles used to classify both individual oligomers and oligomer blends are based on some parameters that characterize structural organization of a system either in the initial (uncured) or final (cured) states. Most of these distinctive parameters, such as, for example, reactivity, number and distribution of functional groups, molecular mass of oligomeric block and its chemical nature, may be used to characterize both individual oligomers and oligomer blends. However, some features are associated exclusively with the structure of blends. These are the structural parameters characteristic of phase organization of a system. Before we proceed to the discussion of these parameters, it is appropriate to outline the general structural hierarchy adopted in polymer chemistry [14]. This is necessitated by the fact that polymer structure is a very general and not very distinctly specified concept and term, which leaves place to superficial scientific speculations and misleading inferences.

1.3. STRUCTURAL HIERARCHY

Two levels of structural organization–molecular and supermolecular–are usually considered when discussing linear polymers [45 - 47].

Molecular level reflects the chemical structure of the elements of a system: chemical composition of repeating units, nature of covalent bonds linking the units into a chain, nature of terminal and side groups, functionality, stereochemical composition (cis, trans, iso, syndio, etc.), arrangement of units in a chain (1-2, 1-4, head-to-tail, ...), etc.

Supermolecular level reflects the spatial arrangement of the elements of a system: nature and strength of physical bonds involved in intermolecular interaction, the extent of order in the arrangement of molecules or their segments, concentration fluctuations (formation of associates, cybotaxes, domains, clusters, etc.—we will discuss the relevant terminology later), and some other parameters.

For the cross-linked polymers, a topological level* of structural organization is additionally considered [14].

Topological level reflects the distribution of network junctions with respect to their number and branching functionality, distribution of interjunction chains in length, distribution of "internal" cycles with respect to their number and size, etc. These characteristics ignore the chemical structure of junctions and cycles and the nature of related bonds. Mathematically polymer topology is described in terms of graph theory [14] or in terms of "bond blocks"—a recently suggested concept that appears to be very fruitful [52].

The three levels described are not sufficient to describe blends. The properties of these systems are primarily controlled by the parameters belonging to colloidal level [11, 29, 30]. For example, a phase transition is always accompanied by a jumpwise variation of macroscopic properties of the system.

Colloidal level reflects the phase and morphological organizations of a system. It is described by phase state diagrams; distribution of dispersed particles with respect to their number and size (dispersity); chemical affinity between the components which controls their equilibrium concentrations in the corresponding phases; chemical nature of the continuous and the dispersed phases; kinetic morphological stability; phase inversion; interfacial phenomena; etc.

All parameters, though belonging to different levels of the structural hierarchy, are interrelated. For example, the structure of repeating unit controls the chemical affinity between the components of a blend, which, in its turn, controls density fluctuations, the packing in supermolecular structures, and equilibrium concentrations of the components in coexisting phases. Another example: functionality of the reactive groups of constituent molecules controls the branching functionality of the resultant network, which, in its turn, controls all parameters relevant to the dispersity of a system. However, the

---

* Some topological characteristics (e.g., MMD) are important for linear products as well.

character of these relationships is far from being lucid. Indeed, it has yet to be established how these relationships are manifested in real processes of oligomer processing, how the kinetic limitations affect the attainment of thermodynamically allowed states, and how far from the thermodynamic equilibrium does the system get frozen, and how does the nonequilibrium state of the system (and the distance from equilibrium) affect the properties of the final material. We believe that these and other problems related to the interplay between thermodynamic and kinetic factors during the processing of oligomer blends to materials must constitute the subject of modern physical chemistry and materials science of oligomer blends.

Note also the essential difference between the oligomer blends based on reactive and nonreactive oligomers. Indeed, in reactive oligomer blends, the formation of topological, supermolecular, and colloidal structures of the cured product (polymer–polymer blend) occurs concomitantly with the permanent variation of the chemical structure of the components (at least of one of them), whereas the formation of materials from nonreactive components (e.g., common blends of two or more polymers) is not accompanied by any chemical transformations*. Hence, the changes in molecular structure resulting from the chemical reactions of an oligomer inevitably lead to rearrangements on all the other levels of structural hierarchy. This means that during the cure of reactive oligomer blends, both the thermodynamic parameters of the system as a whole and those characteristic of each of the components vary. Because of this, the equilibrium, that is, thermodynamically stable values of the parameters belonging to different structural levels vary during the course of the process, whereas in nonreactive blends these parameters are a function of the parameters of state only and are specified from the very start of the process.

---

* In an ideal case, chemical transformations, for example, degradation, must be suppressed during polymer processing.

REFERENCES

1. Gol'danskii, V.I., *et al.*, *Fizicheskay khimiya. Sovremennye problemy* (Physical Chemistry. Modern Aspects), Moscow: Khimiya, 1988 (in Russian).
2. Fox, R., *Energy and the Evolution of Life*, New York: Fridman, 1988.
3. *Kratkay khimicheskaya entsiklopediya* (Concise Chemical Encyclopedia), Moscow: Sov. Entsiklopedia, 1964, vol. 3, p. 725 (in Russian); *Entsiklopedia polimerov* (Polymer Encyclopedia), Moscow: Sov. Entsiklopedia, 1974, vol. 2, p. 457 (in Russian).
4. Berlin, A. A. and Mezhikovskii, S.M., *Zh. Vses. Khim. o-va im. D.I. Mendeleeva*, 1975, no. 2, p. 457.
5. Flory, P., *Statistical Mechanics of Chain Molecules*, New York: Interscience, 1969.
6. Mark, H. and Flory, P., *J. Am. Chem. Soc.*, 1966, vol. 88, p. 3072.
7. Patrikeev, G.A., *Dokl. Akad. Nauk SSSR*, 1975, vol. 221, no. 4, p. 134.
8. Vainshtein, E.F., *Dokl. Akad. Nauk SSSR*, 1976, vol. 229, no. 5, p. 336.
9. Entelis, S.G., Evreinov, V.V., and Kuzaev, A.I., *Reaktsionnosposobnye oligomery* (Reactive Oligomers), Moscow: Khimiya, 1985 (in Russian).
10. Berlin, A.A., *I Vses. konf. po khimii i fizikokhimii polimerizatsionnosposobnykh oligomerov* (Abstracts of Papers, First All-Union Conf. on the Chemistry and Physical Chemistry of Polymerizable Oligomers), Chernogolovka, Akad. Nauk SSSR, 1977, vol. 1, p. 8 (in Russian).
11. Kuleznev, V.N., *Smesi polimerov* (Polymer Blends), Moscow: Khimiya, 1980 (in Russian).
12. Lipatov, Yu.S., *I Vses. konf. po khimii i fizikokhimii polimerizatsionnosposobnykh oligomerov* (Abstracts of Papers, First All-Union Conf. on the Chemistry and Physical Chemistry of Polymerizable Oligomers), Chernogolovka, Akad. Nauk SSSR, 1977, vol. 1, p. 59 (in Russian).
13. Paul, D.R., In: *Polymer Blends*, Paul. D.R. and Newman, S., Eds., New York: Academic, 1978.

14. Irzhak, V.I., Rozenberg, B.A., and Enikolopov, N.S., *Setchatye polimery* (Polymer Networks), Moscow: Nauka, 1979 (in Russian).
15. Sedov, L.N. and Mikhailova, Z.V., *Nenasyshchennye poliefiry* (Unsaturated Polyesters), Moscow: Khimiya, 1977 (in Russian).
16. Omel'chenko, S.I., *Slozhnye oligoefiry i polimery na ikh osnove* (Oligoesters and Related Polymers), Kiev: Naukova Dumka, 1976 (in Russian).
17. Chernin, I.Z., Smekhov, F.M., and Zherdev, Yu.V., *Epoksidnye polimery i kompozitsii* (Epoxide Polymers and Composites), Moscow: Khimiya, 1982 (in Russian).
18. Lee, H. and Neville, K., *Epoxy Resins,* New York: McGraw-Hill, 1957.
19. Saunders, J.H. and Frisch, K.C., *Polyurethanes: Chemistry and Technology.* Part 1, *Chemistry,* 1962; Part 2, *Technology,* 1964.
20. Kercha, Yu.Yu., *Fizicheskaya khimiya poliuretanov* (Physical Chemistry of Polyurethanes), Kiev: Naukova Dumka, 1979 (in Russian).
21. Morozov, Yu.L., *Fizikokhimiya protsessov pererabotki reaktsionnosposobnykh oligomerov v elastomernye izdeliya* (Physical Chemistry of the Processing of Reactive Oligomers to Elastomeric Articles), Chernogolovka: Akad. Nauk SSSR, 1986 (in Russian).
22. Malkin, A.Ya. and Begishev, V.P., *Khimicheskoe formovanie polimerov* (Reaction Molding of Polymers), Moscow: Khimiya, 1991 (in Russian).
23. Sperling, L.H., *Interpenetrating Polymer Networks and Related Materials,* New York: Plenum, 1981.
24. Lipatov, Yu.S. and Sergeeva, L.M., *Vzaimopronikayushchie polimernye setki* (Interpenetrating Polymer Networks), Kiev: Naukova Dumka, 1979 (in Russian).
25. Kozlov, P.V. and Papkov, S.P., *Fiziko-khimicheskie osnovy plastifikatsii polimerov* (Physico-Chemical Principles of Polymer Plasticization), Moscow: Khimiya, 1982 (in Russian).
26. Barshtein, R.S., Kirilovich, V.I., and Nosovskii, Yu.E., *Plastifikatory dlya polimerov* (Polymer Plasticizers), Moscow: Khimiya, 1982 (in Russian).
27. Dontsov, A.A., Kanauzova, A.A., and Litvinova, T.V., *Kauchuk-oligomernye kompozitsii* (Rubber–Oligomer Compositions), Moscow: Khimiya, 1986 (in Russian).
28. Shtarkman, B.P., *Plastifikatsiya polivinilkhlorida* (Plasticization of Polyvinyl Chloride), Moscow: Khimiya, 1986 (in Russian).
29. Lipatov, Yu.S., *Kolloidnaya khimiya polimerov* (Colloid Chemistry of Polymers), Kiev: Naukova Dumka, 1984 (in Russian).
30. Mezhikovskii, S.M., *Nekotorye problemy fizikokhimii polimer-oligomernykh sistem i kompozitov na ikh osnove* (Some Problems in Physical Chemistry of Polymer–Oligomer Systems and Related Composites), Chernogolovka: Akad. Nauk SSSR, 1986 (in Russian).
31. Shaboldin, V.P., *et al., Usp. Khim.,* 1976, no. 45, p. 160.
32. Vinogradov, V.N. and Malkin, A.Ya., *Reologiya polimerov* (Polymer Rheology), Moscow: Khimiya, 1977 (in Russian).
33. Mezhikovskii, S.M., In: *Sinteticheskie oligomery* (Synthetic Oligomers), Perm': Akad. Nauk SSSR (in Russian).

34. Churakova, I.K. and Mezhikovskii, S.M., *Osnovnye napravleniya proizvodstva oligomerov dlya khimicheskogo formovaniya* (Major Trends in Production of Oligomers for Reaction Molding), Moscow: NIITEKhim, 1981 (in Russian).
35. Manson, J.A. and Sperling, L.H., *Polymer Blends and Composites*, New York: Plenum, 1976.
36. Frenkel'. Ya.I., *Kineticheskaya teoriya zhidkosti* (Kinetic Theory of Liquid), Leningrad: Nauka, 1975 (in Russian).
37. Bekturov, E.A., *Troinye polimernye sistemy v rastvorakh* (Ternary Polymer Systems in Solution), Alma-Ata: Nauka, 1975 (in Russian).
38. Petrov, G.N., Kalaus, A.E., and Belov, I.B., In: *Sinteticheskii kauchuk* (Synthetic Rubber), Leningrad: Khimiya, 1976 (in Russian).
39. Churakova, I.K. and Mezhikovskii, S.M., *Khimicheskoe formovanie oligomerov i monomerov* (Reaction Molding of Oligomers and Monomers), Moscow: NIITEKhim, 1979 (in Russian).
40. Mogilevich, M.M., Turov, B.S., Morozov, Yu.L., and *Ustavshchikov, Zhidkie uglevodorodnye kauchuki* (Liquid Hydrocarbon Rubbers), Moscow: Khimiya, 1983 (in Russian).
41. Mezhikovskii, S.M., *Oligomery* (Oligomers), Moscow: Znanie, 1983 (in Russian).
42. Rozenberg, B.A., In: *Sinteticheskie oligomery* (Synthetic Oligomers), Perm': Akad. Nauk SSSR, 1988, p. 68 (in Russian).
43. Barantsevich, E.N. and Time, T.A., In: *Kauchuk-89. Problemy razvitiya nauki i proizvodstva* (Rubber-89. Problems in Science and Production), Moscow: TSNIITENeftekhim, 1990, vol. 1, p. 78 (in Russian).
44. Baturin, S.M., In: *Sinteticheskie oligomery* (Synthetic Oligomers), Perm': Akad. Nauk SSSR, 1988, p. 44 (in Russian).
45. Kargin, V.A. and Slonimskii, G.L., *Kratkie ocherki po fizikokhimii polimerov* (Sketches in Physical Chemistry of Polymers), Moscow: Khimiya, 1967 (in Russian).
46. Yeh, G.S., *Vysokomol. Soedin.*, Ser. A., 1979, vol. 21, no. 11, p. 2433.
47. Amerik, Yu.B. and Krentsel', B.A., In: *Uspekhi khimii i fiziki polimerov* (Advances in Polymer Chemistry and Physics), Moscow, Khimiya, 1973 (in Russian).
48. *Reaction Injection Molding and Fast Polymerization Reactions*, Kresta, I.E., Ed., New York: Plenum, 1982.
49. *Reaction Injection Molding*, Kresta, I.E., Ed., Washington, ACS, 1985.
50. Entelis, S.G., *III Vses. konf. po khimii i fizikokhimii oligomerov* (Abstracts of Papers, Third All-Union Conf. on Chemistry and Physical Chemistry of Oligomers), Chernogolovka: Akad. Nauk SSSR, 1986, p. 11 (in Russian).
51. Baturin, S.M., *V Vses. konf. po khimii i fizikokhimii oligomerov* (Abstracts of Papers, Fifth All-Union Conf. on Chemistry and Physical Chemistry of Oligomers), Chernogolovka: Akad. Nauk SSSR, 1994, p. 4 (in Russian).
52. Irzhak, V.I., Tai, M.L., Peregudov, N.I., and Irzhak, T.E., *Colloid Polym. Sci.*, 1994, vol. 272, p. 523.

2   THERMODYNAMICS OF OLIGOMER BLENDS

2.1. COMPATIBILITY AND PHASE EQUILIBRIA IN OLIGOMER SYSTEMS

In the "Encyclopedia of Polymers" [1] the "compatibility of polymers" is determined as "a term that is commonly accepted in technological practice, which characterizes the ability of various polymers to form mixtures possessing satisfactory properties". Definitions with analogous meaning are also given in monographs [2, 3]. Not dwelling here on the reasons for which these definitions cannot be considered as correct from the standpoint of physics, note only that products of oligomer conversion, rather than oligomer blends in the initial state, are usually employed as materials or articles. Therefore, it is obvious that the notion of compatibility, in the sense implied in polymers, cannot be applied to oligomer systems.*

---

* For some time, the term "compatibility" was used in the literature on polymers in conjunction with adjectives such as "technological", "mechanical", "induced", etc. In our opinion, this differentiation has no strict physical meaning. The use of these terms is a consequence of unsuccessful attempts to attach a scientific significance to technological methods used in obtaining polymer blends with "satisfactory" (?) technological and mechanical properties, or to analyze the phase separation ignoring the kinetic factors.

By the compatibility of oligomer blends we will imply, following [4–7], a complete mutual solubility of components in the mixture resulting from their molecular dispersity or, in other words, the ability of components to form true solutions. The same pair of components, constituting an oligomer blend, can be either compatible or incompatible, depending on the temperature and the ratio of components (i.e., on the parameters of state). In the former case, a dispersion is formed on the molecular level (representing a true solution or a single-phase system). In the latter case, the components either exhibit no dissolution at all (showing even no traces of swelling), which is virtually never observed in practice, or form a two-phase colloidal system comprising a dispersion of two co-existing phases, each one of these being a true solution enriched with or depleted of the oligomer. The equilibrium concentrations of components in the co-existing phases are determined by the phase diagram of the system.

The compatibility of oligomer blends has at least four interrelated aspects: thermodynamic, kinetic, colloidal, and methodological. The thermodynamic aspect is determined by applicability of the fundamental rule of phase equilibrium and the methods of statistical thermodynamics to the system studied. The kinetic aspect is characterized by the time of attaining the state of thermodynamic equilibrium. The colloidal factor reflects the stability of two-phase systems involved and the role of interfaces in the manifestations of macroscopic properties of the given colloidal system. The methodological aspect is related to the discrepancy of information, obtained by various methods of determination of the phase composition, and the ambiguity of the concept of "phase" as employed by various researchers. Below we will consider these aspects in some detail and formulate several problems of importance in practical applications.

Thermodynamic Aspect

As long ago as in 1937, V.A. Kargin, S.P. Papkov, and Z.A. Rogovin showed that polymer solutions obey the rule of phase equilibria, which is one of the fundamental laws of physical chemistry. Since then, there were numerous attempts to find an analytical relation between the molecular parameters of components and their mutual solubilities. However, no serious generalizations were obtained for a long time. It was only the lattice model of liquid created by Flory and Huggins [5] using the concepts of the Gibbs equilibrium thermodynamics, which analytically related a change in the free energy of mixing (ΔG_m) or a change in the chemical potential ($\Delta \mu_i$) to the equilibrium fraction of polymer in solution (φ_2). Then the theory was developed, so as to apply to polymer blends, by Scott and Tompa (see [6–8]) to yield an equation relating ΔG_m to the molecular masses of components and the dissolution temperature:

$$\Delta G_m = \frac{RTV}{V_s} \left[\frac{\varphi_1}{x_1} \ln \varphi_1 + \frac{\varphi_2}{\varphi_1} \ln \varphi_2 + \chi \varphi_1 \varphi_2 \right], \qquad (2.1)$$

where V is the total volume of the mixture, V_s is the molar volume of the monomer unit (assumed equal for two components of the mixture), φ_1 and φ_2 are the volume fractions of mixed components, X_1 is the degree of polymerization, and χ is the parameter of interaction, or the Flory–Huggins constant, which characterizes the excess free energy of interaction per solvent molecule in the polymer solution.

By equating to zero the second and third derivatives of ΔG_m with respect to the volume, the following relations are obtained for the critical values φ_{cr} and χ_{cr}:

$$\varphi_{1cr} = 1 - \varphi_{2cr} = \frac{1}{\sqrt{1 - (x_2/x_1)^{1/2}}} \qquad (2.2)$$

$$\chi_{cr} = \frac{1}{2} \left[\left(\frac{1}{x_1} \right)^{1/2} + \left(\frac{1}{x_2} \right)^{1/2} \right]^2 \qquad (2.3)$$

where $X_1 = V_1/V_0'$, $X_2 = V_2/V_0''$, V_1 and V_2 are the molar volumes of the components, V_0' and V_0'' are the reference coefficients equal to the molar volumes of the corresponding monomer units.

In this approximation, the Scott theory leads to the value $\chi_{cr} = 2$ for the case of mixing two low-molecular-mass liquids (for which $X_1 \approx 1$ and $X_2 \approx 1$). For a mixture comprising a polymer ($X_2 \to \infty$) and a low-molecular-mass liquid ($X_1 = 1$), the theory yields $\chi_{cr} = 0.5$, and for a mixture of two oligomers ($X_1 \approx 10^3$ and $X_2 \approx 10^3$) we have $\chi_{cr} = 0.01$. At the same time, the theory predicts incompatibility of components in a mixture of two polymers ($X_1, X_2 \to \infty$). Thus, it might seem that compatibility of any pair of components could be judged by their χ_{cr} values. However, it was found that experiment was far from coincidence with the theory. The most striking example of disagreement, which cast doubt on the principal concepts of the theory, was that several pairs of compatible (i.e., forming true solutions) polymers were known in contrast to the theoretical prediction.

In order to remove this and some other similar uncertainties, various "refinements" were introduced into the theory, which were not contained in the initial aprioric concepts. In particular, the "segmental" solubility was explained by Macknight et al. [6] with the aid of an empirical parameter S_1 describing the so-called "degree of mixing". The dispersion of components on the molecular level corresponds to $S_1 = 1$. The case of $S_1 < 0$ reflects the dissolution occurring on the level of segments, and $S_1 \gg 1$ describes the phase separation.

These ways of fitting the theory to experiment were rather artificial. The disadvantages of the classical theory, which had been inherited from restrictions of the Hildenbrandt lattice model, were later removed by Flory in the refined thermodynamic theory of liquid mixtures. Not going into detail of this theory, we will only mention that the refined version quite naturally took into account some specific features of real polymer solutions, including a change in the free volume upon mixing, association of components, polydispersity of polymers with respect to molecular mass, etc.

An essential feature of the theory as refined by Flory consisted in showing that the mixing of polymers may produce an increase in the degree of association of molecules for both component polymers, whereby the total entropy of the system would decrease at the expense of contribution due to the non-combinatory component of the entropy term of the free energy.

Analytical equations of the refined theory contain an interaction parameter χ_{12} which includes, like its classical analog, the enthalpy (χ_H) and the entropy (χ_S) components, but has a differing physical meaning, representing a reduced chemical potential:*

$$\chi_{12} = \chi_H + \chi_S = \frac{\Delta\mu_1^R}{RT\varphi_2^2} = \frac{\Delta H_1}{RT\varphi_2^2} - \frac{\Delta S_1^R}{RT\varphi_2^2} = \frac{\Delta\mu_1 - T\Delta\bar{S}_1^K}{RT\varphi_2^2}, \quad (2.4)$$

where ΔS_1^C and ΔS_1^R are the combinatory and non-combinatory (residual) entropies of mixing, respectively, ΔH_1 is the enthalpy of mixing, $\Delta\mu_1^R$ is the residual chemical potential, and $\Delta\mu_1$ is the change of the chemical potential.

The refined Flory theory allowed several unusual results, which appeared anomalous at that time, to be explained, including a decrease in the system entropy upon mixing the components, the exothermal effect observed upon mixing some incompatible polymers, etc. However, there are effects not yet explained within the framework of the existing equilibrium thermodynamic theories [3, 7–12, 59, 84].

---

* If polymers are mixed in a common solvent, this parameter is denoted by χ_{23}.

Kinetic Aspect

High viscosities of polymers and differing velocities of mutual diffusion of the components from various phases in a mixture are factors determining the difficulty of attaining the equilibrium state of such systems. Therefore, the methods of sample preparation for testing and the conditions and duration of sample storage (i.e., the prehistory determining the degree of approaching the equilibrium) may affect the experimental results. This very circumstance may account for the highly contradictory experimental data reported on the mutual solubilities of components in polymer blends [2, 3, 10].

This is well illustrated by the results reported by Kwei *et al.* [13], where the films of a mixture comprising 70% oligovinylmethyl ether and 30% polystyrene, obtained from solution in a common solvent, were transparent for some time after vacuum drying (solvent removal), but got turbid when stored for 1.5–2 months. Similar films containing 75% ether became turbid already in 5–8 days. Since no chemical processes were shown to occur during the storage of both compositions, the initial transparency of the films cannot serve as indication of the thermodynamic compatibility of components. Indeed, if the optical densities of both films are measured in the first hours or days after preparation, we may incorrectly conclude on the single-phase character of the systems. The given system is incompatible, but the phase separation proceeds at a very low rate, which results from high viscosity of the components. In this regard, it is worth recalling a remark made, albeit in some other connection, by Prof. V.I. Irzhak: "Thermodynamics provides, kinetics decides". Thermodynamics determines the potential, limiting equilibrium states of a system, while the kinetics controls the time (which sometimes becomes infinitely long as compared to the time of observation) required for the system to attain the equilibrium. Unfortunately, no unique criteria have been reported in the literature that can be used to unambiguously judge as to whether a given mixture has attained the equilibrium state. We will proceed with discussing this problem below.

Colloidal Aspect

In considering the process of cellulose acetate dissolution in chloroform and the subsequent phase organization in this system, Papkov [9] was the first to distinguish several stages in the phase state of polymer systems with limited compatibility. First, as the temperature is decreased, the solution passes a critical point and the system shows turbidity (*phase separation*), suggesting that an emulsion is formed. Then, particles of like phases gradually merge together to form two transparent layers, representing a cellulose solution in chloroform (top layer) and a chloroform solution in polymer (bottom layer). The latter process represents the *phase delamination*. Similar observations were also reported in other systems, including the oligomer blends. In the aforementioned system of cellulose acetate–chloroform, the phase separation is a long-term process with the rate depending on the drop size, the difference between the densities and viscosities of phases involved, the magnitude of surface tension, the presence of surfactants, etc.

In heterogeneous oligomer–oligomer (and the more so, polymer–oligomer) systems, where the viscosity of a polymer-rich phase exceeds by 10–12 orders of magnitude that of the second phase, the process of macroscopic delamination may extend over a period of time markedly exceeding any reasonable time acceptable for a real technological cycle. A process engineer must take this into account and analyze the phase delamination and morphology formation in oligomer blends using the developments of classical physicochemical mechanics [14] and the colloid chemistry of polymers [15]. We will return to this problem in Section 2.8.

Methodological Aspect

The variety of experimental methods employed in the study of thermodynamic compatibility of components of blended systems is based on four principles: (1) measurement of the optical properties of the system; (2) determination of the sorption capacity; (3) registration of the relaxation characteristics of the blend; (4) monitoring of the properties of a third component (label) introduced into the system. Accordingly, the experimental techniques can be also divided into four groups with the corresponding instrumentation [16].

Each method within a given group has advantages and disadvantages of its own. There are objective restrictions on the applicability of various techniques. For example, measurements of the optical density of a system cannot, in principle, provide information about particles with dimensions below 0.1 μm, which are therefore not detected even if present in the system. On the other hand, methods based on the X-rays (e.g., the small-angle X-ray scattering) cannot "see" particles greater than 0.1 μm. In accordance with the laws of optics, we cannot differentiate between phase formations if the dispersion medium and the dispersed phase have close indices of refraction. Therefore, even obviously heterogeneous systems may appear as transparent in optical experiments.

Paramagnetic probes are characterized by rather strict limits with respect to the correlation times, above and below which variations in the system mobility cannot be detected. An objective limitation of the methods of dynamic testing is the impossibility to study the phase structure of a system in a wide temperature interval. Indeed, if a sample to be tested by mechanical or dielectric spectroscopic techniques, differential thermal analysis, radiation- and thermo-induced luminescence, etc., is prepared at room temperature and then rapidly quenched (in accordance with the experimental procedure employed), the temperature-induced transitions revealed upon the subsequent testing would correspond only to the state and structure that are developed in the sample under the particular preparation conditions and then "fixed" by suddenly decreasing the temperature. Moreover, variations in the method of component mixing (by dissolving in a common solvent or by swelling, mechanical dispersion, etc.) may increase the uncertainties and lead to errors caused by the kinetic limitations considered above.

Differences in the "sensitivity" of various methods and the procedures used for the sample preparation account for the contradictory data on the phase organization of blends, which are sometimes obtained by various techniques. This is well illustrated by the work of Macknight et al. [6], where a mixture of poly(dimethylphenylene oxide) with polystyrene was systematically studied by a number of methods, including the dynamic mechanical spectroscopy (loss modulus E''), static mechanical tests (the Young modulus and elastic modulus E'), determination of dielectric losses, and thermooptical measurements. Some of these techniques detected a single glass transition temperature T_g in the blend studied, which corresponded to a weight-average T_g value for the homopolymers and was indicative of the compatibility of components. At the same time, other methods revealed two T_g values close to those of the individual components, which showed evidence that the mixture was incompatible. Moreover, one of the methods showed a curve with three peaks, from which we must infer that the system had three glass transition temperatures (the weight-average and two individual) and exhibited a limited compatibility. Thus, the same system studied by various methods can be assigned different phase organizations, which must be excluded by critical analysis of the possibilities of various techniques and by correct processing of the experimental data.

Another typical mistake to be noted is most frequently encountered in experiments using optical microscopy. Some researchers are prone to interpret any feature displayed by a microscope in terms of the phase formation. However, the presence of microinhomogeneities in the area of observation (assuming artifacts excluded) cannot be treated as unambiguous indication of the presence of second phase in the mixture. In fact, even a single-phase blend can be inhomogeneous. It is important to decide whether a phase transition in the system took place or not, but this problem is not always correctly considered. According to the classical definition formulated by Gibbs [69], a phase is a unity of homogeneous parts of a system with like composition, possessing chemical and physical properties independent of the amount of substance, which are separated by an interface from other parts of the system. Even taking into account, following the remark of A.V. Rakovskii, that the class of thermodynamic properties includes those (and only those) of the physical properties that can be represented as functions of the parameters of state (i.e., of the volume, pressure, temperature, concentration) and only of these parameters [17],* the phase is yet characterized by a combination of two significant qualities: (i) independence of these properties of the parameters of state and (ii) the presence of interface. The combination of these qualities is the necessary and sufficient condition for the existence of phases. It is only the combination of the two qualities that allows us distinguishing thermodynamic phases from inhomogeneities of other types.

The dimensions of macromolecular coils, not speaking of the molecular aggregates and fluctuating cybotaxic formations, may well reach the micron level. During microscopic observations, these objects may be incorrectly interpreted as particles of a second phase. The confusion is related to the fact that, in one case, the physical and chemical properties of inhomogeneities, bounded by interfaces, depend on the amount of substance in the system, while in the other case, the inhomogeneities, whose thermodynamic functions are independent of the parameters of state, possess no clear phase boundaries.

2.2. PHASE DIAGRAMS OF OLIGOMER BLENDS

Among the components employed in practically significant oligomer blend systems, there are only a few that can be considered as com-

---

* According to A.V. Rakovskii, some physical properties, including electrical, magnetic, crystallographic, etc., do not belong to the class of thermodynamic functions. Later, Fischer developed this approach by subdividing thermodynamic quantities into the "field" and "density" variables.

pletely compatible (i.e., compatible within the entire range of compositions and temperatures close to those encountered in real technological processes). These are the aforementioned monomer–oligomer mixtures of styrene with oligomaleate fumarate and amines with epoxy oligomers, the polymer–oligomer mixtures of butadiene-nitrile rubber (SKN-40) with trioxyethylene dimethacrylate, all oligomer–oligomer mixtures of oligobutadieneurethane methacrylates with oligomers of the n-methylene dimethacrylate series ($n = 1$–10), and some others.

Most of the oligomer blends are systems with limited compatibility, forming true solutions only within a restricted domain of compositions and temperatures. Modern technologies employ a wide range of compositions, which may be treated within temperature intervals of large amplitude even in a single technological process. Therefore, phase transitions are not excluded and it is necessary for a process engineer to know the phase organization of the system in all stages of its evolution.

Quantitative information about the limits of mutual solubility of components in systems with limited compatibility is contained in phase diagrams. It should be noted that the compatibility of components in oligomer blends has been studied for more than four decades. However, some data (in particular, the early ones) must be treated with care for the reasons considered in the preceding section. Complete phase diagrams and reliable experimental values of χ_{12}, ΔH_1, $\Delta\mu_1^R$ and other thermodynamic parameters were obtained only during the last decade, and only for a comparatively small number of oligomer systems. The numerical values of limiting mutual solubilities and data on the phase organization in particular double oligomer blends can be found in monographs [7, 18, 19], where numerous experimental results are summarized.

Below we will present, for illustrative purposes, some of the phase diagrams reported in the literature. In order to analyze the information contained in these diagrams, we will use a schematic phase diagram (Fig. 2.1) constructed in the coordinates of temperature (T) versus the average content of oligomer introduced into the system (c) on the basis of generalization of the experimental data available.

In Fig. 2.1, the domains of single-phase state of the system (I and II) are determined by a binodal (solid curve). The binodal and the domain of two-phase state of the mixture (region V), which is bounded by a spinodal (dashed curve), are separated by a domain (region III) corresponding to metastable states of the system. In this domain, in contrast to the stable single-phase and two-phase states, the system is homogeneous but exhibits phase transitions for any change in the parameters of state (ΔT or Δc). Therefore, even insignificant variations

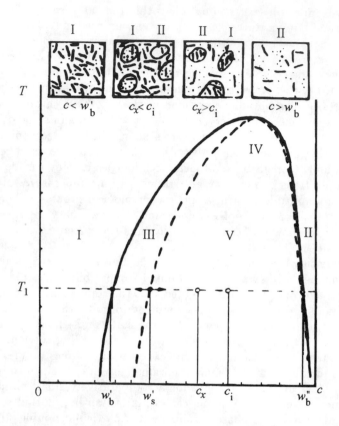

Figure 2.1. Schematic phase diagram of a polymer–oligomer system. Insets show diagrams of the structural organization of single-phase solutions I and II and two-phase blends before and after the point of phase inversion: solid bars, polymer; dots, oligomer. See the text for explanations.

of the temperature or the oligomer content (it is important that ΔT, $\Delta c \neq 0$) result in that a system occurring in the metastable domain instantaneously "jumps" into one of the stable domains, where it becomes a stable single-phase or two-phase system. The microstructure of a mixture occurring in a metastable state, like that of a mixture in the vicinity of a critical point (region IV), differs from the structure of other states by much higher amount and magnitude of fluctuations of the concentration and density of supermolecular formations, that is, by greater contents and longer lifetimes of associates and cybotaxic formations.

Typical features in phase diagrams of the most of oligomer blend systems are the asymmetry and the presence of upper critical solution temperature (UCST), which implies that the mutual solubility of components always increases with the temperature. This is manifested by the limiting solubility of oligomer in the second component (domain I) markedly exceeding the concentration corresponding to the equilibrium solubility of second component in the oligomer (domain II).*

As is seen from Fig. 2.1, the system remains single-phase at a given temperature T_1 (e.g., a temperature of blend storage) for the average oligomer contents $c < w_b'$, where w_b' is the concentration corresponding to the point of intersection of the line $T = T_1$ with the left-hand branch of binodal. These states are characterized by the formation of a true solution of the oligomer in the second component (also called an oligomer-depleted solution). For the oligomer contents $c > w_b''$, where w_b'' is the concentration corresponding to the point of intersection of the line $T = T_1$ with the right-hand branch of binodal, the system also has a single-phase character, but comprises a solution of the second component in the oligomer (also called an oligomer-rich solution). As the average oligomer content increases to cross the left-hand branch of binodal ($c > w_b'$), the system falls within the metastable region III extended to the point $c = w_s'$, which corresponds to the oligomer concentration on the left-hand branch of spinodal. The asymmetry of phase diagrams of the oligomer blend systems accounts for the fact that the right-hand branches of spinodal and binodal virtually coincide ($w_s'' \approx w_b''$). On passage to the region $c > w_s'$, the system becomes two-phase. It should be emphasized that the two-phase blend is not a mixture of the initial oligomer and second component separated by a phase boundary. The interval $w_s' < c < w_b''$ represents a mixture of true solutions I and II with the oligomer concentrations w_b' and w_b'', respectively (see insets in the top part of Fig. 2.1).

Another important detail to be noted is that in single-phase systems the average content of oligomer c (i.e., the average content introduced into the system) always coincides with its concentration ($w \equiv c$), while in two-phase systems the two quantities are generally different ($w \neq c$). In the latter case, the concentrations of oligomer in

---

* The second component of an oligomer blend system can be a monomer, oligomer, or polymer.

the coexisting phases are always constant (w_b' and w_b''). An increase in the total amount of oligomer in the two-phase blend at $T = \text{const}$ does not lead to a change in the concentrations of component solutions, but alters the volume ratio of dispersed and continuous phases and results eventually in the phase inversion. This situation is also illustrated in Fig. 2.1, where the continuous phase is represented by solution I at $c < c_i$, and by solution II at $c > c_i$.

Thus, once the phase diagram of a particular oligomer blend system is known, a process engineer may vary the composition and temperature in order to preset a desired phase organization in the system.

2.3. INFLUENCE OF THE MOLECULAR PARAMETERS OF COMPONENTS ON COMPATIBILITY IN OLIGOMER BLENDS

Figures 2.2 to 2.4 show typical phase diagrams for systems representing oligomer blends of various kinds. As is seen, all diagrams exhibit UCSTs, irrespective of the chemical nature of components, their molecular masses, and the character of terminal groups. However, the position of binodal is determined by the molecular parameters of an oligomer and the second component in the mixture. Below we will analyze the effect of some parameters on the compatibility of blend components.

Effect of the Molecular Weights of Components and the Nature of Oligomer Block

A comparative analysis of phase diagrams of the oligomer blends representing three homologous series of dimethacrylic esters of n-hydroxyethylene glycols, n-hydroxyethylene(phthalate) glycols, and n-alkanediols with various polymers showed that increasing n (which is equivalent to increasing molecular mass of the oligomer) in the first two systems (including blends with all polymers studied by now) produces a systematic decrease in compatibility.* In the third case, increasing n leads to complicated unusual variation of the compatibility, depending on the particular polymer involved in the blend system.

---

* According to the existing thermodynamic theories of polymer solutions, an increase in the molecular weights of components must lead, through variation of the entropy factor, to a decrease in solubility, as was repeatedly confirmed in experiment [7, 10, 18, 19].

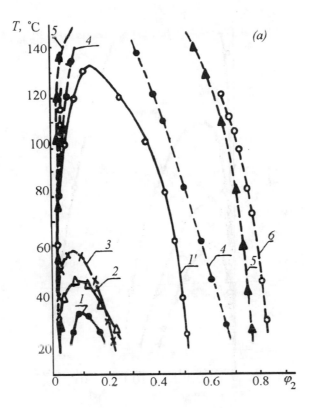

Figure 2.2a. Phase diagrams of oligomer–polymer systems:(*a*) mixtures between *n*-hydroxyethylene dimethacrylates with *n* = 1 (*4*), 3 (*1-3*, *5*), 13 (*1'*, *6*) and butadiene–acrylonitrile copolymers containing 40% (*1*, *1'*), 26% (*2*), and 18% (*3*) acrylonitrile groups (rubbers of the SKN-40, SKN-26, and SKN-18 grades, respectively); (*b*) mixtures between *n*-methylene dimethacrylates (*n* values are indicated at the curves) and an SKN-40 rubber; φ_2 denotes the volume fraction of polymer [20, 21].

This is clearly demonstrated in Fig. 2.5, where the absolute UCST value and its relative change $\gamma = (\text{UCST} - n)/(\text{UCST} - 1)$ are plotted versus the number n of repeated groups in the oligomer block.

Of special interest is the behavior observed in mixtures comprising oligomers of the *n*-methylene dimethacrylate series and *cis*-polyiso-

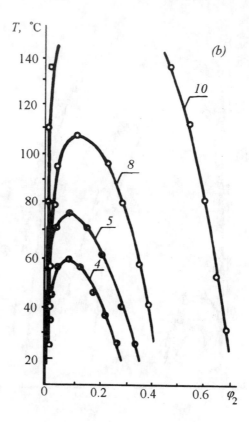

Figure 2.2b.

prene. In contrast to all the known experimental facts and theoretical predictions (see formulas (2.1) to (2.3)), according to which the compatibility of components decreases with increasing molecular masses, this system exhibits increasing mutual solubility of components with growing n value.

Effect of the Polarity of Components

The experimental fact mentioned above, despite its "anomalous" character, is consistent with the role belonging to the polarity of components in the dissolution of polymers. As is known [3], the smaller the difference between the polarities of mixed components, the higher is

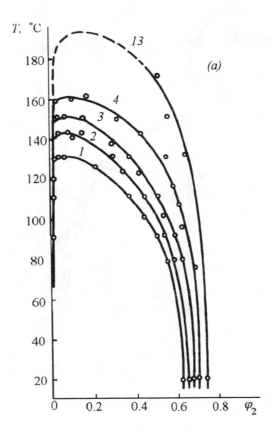

Figure 2.3a. Phase diagrams of mixtures between poly(vinyl chloride) and (*a*) *n*-hydroxyethylene dimethacrylates or (*b*) *n*-ethylene dimethacrylates (*n* values are indicated at the curves); φ_2 denotes the volume fraction of polymer [20, 21].

their mutual solubility (in accordance with the principle that "similar dissolves in similar"). In the above "anomalous" system, a polar oligoester acrylate is introduced into a nonpolar *cis*-polyisoprene. Moreover, the polarity of rubber is constant, while the polarity of oligomer varies with *n*. Indeed, the molecules of *n*-methylene dimethacrylates are composed of fragments of two types: a nonpolar methylene block ($P = 55.7$) and methacrylic ends, linked by a polar ester bond to

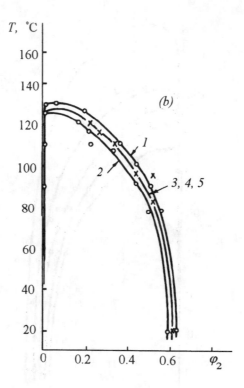

Figure 2.3b.

the oligomer block ($P = 319.4$). The ambipolar character of oligomer molecules is just what accounts for the unusual effect. An increase of n in this homologous series is accompanied by decreasing molar polarity of the oligomer, which must facilitate solubility in the nonpolar *cis*-polyisoprene. However, this is only one side of the problem. An opposite trend consists in that increasing n leads to a growth of the molecular mass, thus detrimentally affecting the solubility. Competition of the two factors determines the compatibility, which is likely to exhibit an extremum when n is varied in a wide interval.

The latter assumption agrees with the results derived both from the classical Flory–Huggins theory of solutions and from the refined Flory theory. Within the framework of the classical theory, with an allowance for the group contributions according to Van Crevelen, Cha-

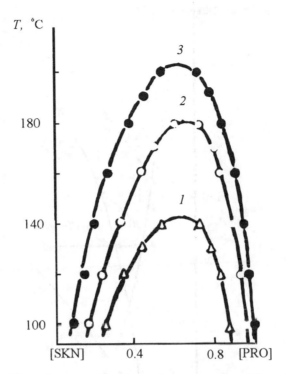

Figure 2.4. Phase diagrams of mixtures between a phenol-formaldehyde oligomer (PFO) and butadiene-acrylonitrile rubbers (*1*) SKN-18, (*2*) SKN-26, and (*3*) SKN-40 [19].

lykh *et al.* [20] obtained an expression relating χ and n to the UCST and the critical concentration of oligomer (w'_{cr}), which allowed the critical characteristics of a mixture between *cis*-polyisoprene and *n*-methylene dimethacrylates to be calculated for n varied from 2 to 10. A comparison of the theory and experiment showed good coincidence of the results (Table 2.1).

An analysis performed within the refined Flory theory allowed contributions of the entropy and enthalpy components of the interaction parameter to the χ_{12} value to be calculated. Such calculations were performed for the aforementioned "anomalous" system with variable n value [21]. The results are presented in Fig. 2.6. A decrease of χ_{12} with increasing n corresponds to the growth of the mutual sol-

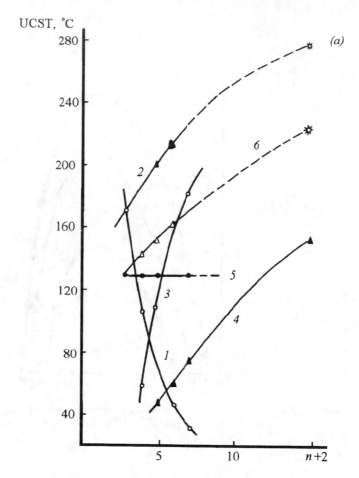

Figure 2.5a. The plots of (*a*) UCST and (*b*) γ versus *n* for the mixtures of (*1, 3, 5*) *n*-ethylene dimethacrylates and (*2, 4, 6*) *n*-hydroxyethylene dimethacrylates with (*1, 2*) SKI-3 *cis*-polyisoprene, (*3, 4*) SKN-40 butadiene-acrylonitrile rubber, and (*5, 6*) poly (vinyl chloride); symbols (*) indicate the calculated data.

ubility of components with increasing molecular masses, which was observed in experiment. However, the contributions of the enthalpy (χ_H) and entropy (χ_S) terms to the change of the free energy of mixing are differing: χ_H drops with increasing *n*, which indicates that the change of enthalpy upon mixing favors the compatibility of com-

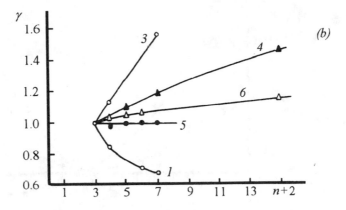

Figure 2.5b.

Table 2.1. Characteristics of the critical state of the *cis*-polyisoprene-*n*-methylene dimethacrylate system

n	UCST, °C		w_{cr}	
	Calculation	Experiment	Calculation	Experiment
2	173	–	0.05	0.06
4	106	108	0.06	0.08
5	51	53	0.07	0.08
8	46	48	0.11	0.12
10	35	35	–	–

ponents, whereas an increase of χ_S suggests that the entropy factor hinders the mutual solubility. Initially, the χ_H contribution prevails over that of χ_S, and the solubility of oligomer in rubber increases with growing molecular mass of the oligomer. For some n values, the χ_H and χ_S contributions compensate one another; as a result, χ_{12} and, hence, the compatibility of components are weakly dependent of the molecular mass of oligomer. The further increase in the number of methylene units in the oligomer molecule results in that the entropy factor begins to dominate and, beginning with $n = 13$–15, the χ_{12} value increases and the compatibility of components decreases.

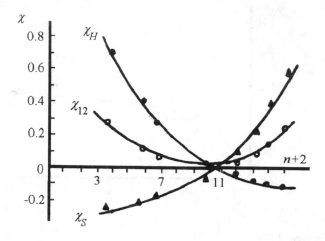

Figure 2.6. The plots of (*1*) χ_{12}, (*2*) χ_H, and (*3*) χ_S versus n for a *cis*-polyisoprene–n-methylene dimethacrylate system at 40°C.

2.4. ANALYSIS OF THERMODYNAMIC PARAMETERS

Figure 2.7 shows the temperature variation of the increment of chemical potential, as calculated using the modified equation (2.4) of the refined Flory theory for the mixtures of poly(vinyl chloride) with oligomers of two homologous series, n-ethylene and n-hydroxyethylene dimethacrylates. The calculation was based on the experimental data reported by Kotova *et al.* [22].

The first fact to be noted is that the character of the temperature dependence of $\Delta\mu_1^R$ exhibits significant change when the system transforms from glassy to rubberlike state. In the mixtures of all oligomers of the n-ethylene series and the first members ($n < 3$) of the n-hydroxyethylene series, the transformation from glassy to rubberlike state affects only the rate of decrease of the $\Delta\mu_1^R$ value with increasing temperature: the compatibility always increases with the temperature, the effect being more pronounced in the rubberlike state than in the glassy state. At the same time, in the mixtures of poly(vinyl chloride) and n-hydroxyethylene dimethacrylates with $n > 3$ the $\Delta\mu_1^R$ value exhibits an increase with growing T in the region below T_g, followed by a drop upon the system transition to

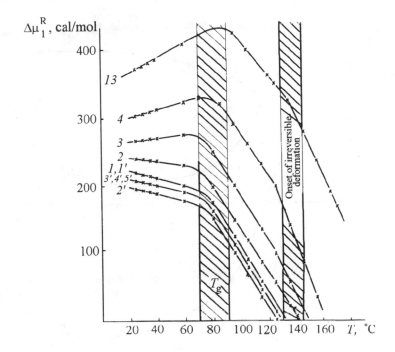

Figure 2.7. Temperature variation of $\Delta\mu_1^R$ for the poly(vinyl chloride) mixtures with (*1-4, 13*) *n*-HEDA and (*1'-5'*) *n*-EDA. Numbers at the curves indicate the *n* values.

the rubberlike state. Therefore, the latter systems would be characterized by a lower critical solution temperature (LCST) if the phase diagrams were restricted to the region of glass transition. However, the tests were performed in a much wider temperature range, and we observe how the growth of $\Delta\mu_1^R$ is changed by the drop. Accordingly, the rubberlike state of these systems is characterized by UCST.

An analysis of changes in the thermodynamic parameters as functions of *n* for the *n*-hydroxyethylene dimethacrylate series confirmed the above-mentioned role of the physical (relaxational) state of the system in manifestation of the affinity of components. As is seen from Figs. 2.8 and 2.9, the $\Delta\mu_1^R$ value increases with the molecular mass, but the contributions of enthalpy and entropy terms to the increment

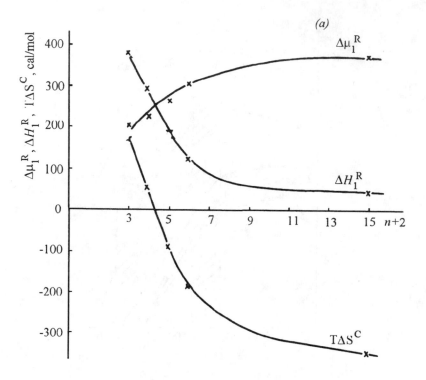

Figure 2.8a. The plots of $\Delta\mu_1^R$, ΔH_1^R, $T\Delta S^C$ (combinatory), and ΔS^R (non-combinatory) versus n for PVC mixtures with n-HEDA at 40°C.

of chemical potential may vary depending on the temperature (above or below T_g) of the blend system.

If the experiments are performed below the point of α-transition (e.g., at 40°C in Fig. 2.8), the growth of $\Delta\mu_1^R$ as a function of n is determined by the fact that a decrease in the non-combinatory entropy term (detrimental for the compatibility) dominates over the decrease in the enthalpy term (favoring the compatibility) in contributing to the free energy of the system. In the temperature interval corresponding to the rubberlike state (e.g., at 100°C in Fig. 2.9), the growth of $\Delta\mu_1^R$ with increasing molecular weight is due to the entropy-related transformations during mixing, because the enthalpy of mixing exhibits no changes upon the variation of n.

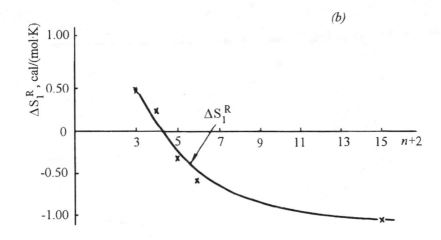

Figure 2.8b.

The observed pattern of variation of the thermodynamic parameters as functions of T and n for the mixtures of poly(vinyl chloride)
and oligomers of n-hydroxyethylene dimethacrylic series is still not
given a strict physical explanation; nevertheless, the laws themselves
and their implementations are of principal significance.*

First, it is important for the practical applications to know that
the temperature variation of compatibility has different trends below
and above T_g. Secondly, a decrease in the non-combinatory entropy
upon mixing the components gives an unambiguous evidence that
mixtures are characterized by increasing degree of the short-range
order as compared to that of the initial components.

The first consequence implies that, in case when the temperature
variations are employed by a process engineer as a means for controlling the compatibility of blend components, the compatibility is
improved by decreasing the temperature if the system occurs in the
state below T_g, and, on the contrary, by increasing the temperature
if the system is above T_g. Of course, no generalizations are possible:
the above recommendation refers only to some particular systems.

---

* In PVC mixtures with oligomers of another homologous series, n-
methylene dimethacrylates, the behavior of thermodynamic parameters was
virtually independent of n.

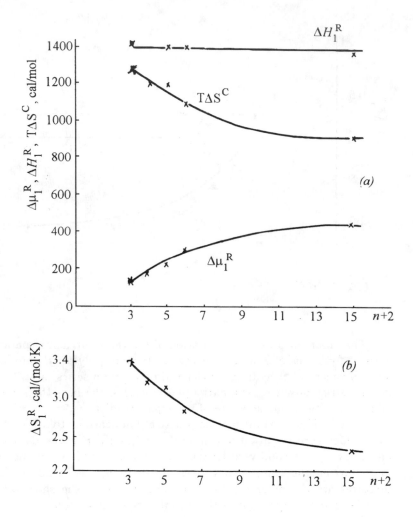

Figure 2.9. The plots of $\Delta\mu_1^R$, ΔH_1^R, $T\Delta S^C$ (combinatory), and ΔS^R (non-combinatory) versus n for PVC mixtures with n-hydroxyethylene dimethacrylates at 100°C.

However, this nontrivial practical conclusion yet confirms that the knowledge of physicochemical laws, and, in particular, of the character of variation of the thermodynamic functions, is of considerable importance for correctly selecting the technological regimes of mixing.

The second consequence gives an experimental evidence for the assumption (which is far from being evident) that even single-phase oligomer blends may have inhomogeneous structures.

Before proceeding with discussion of the inhomogeneous structure of oligomer blend solutions, we must consider another important feature of their phase diagrams. This feature was recently established [85, 86] in the phase diagrams of mixtures using homologous series of oligomers as solvents.

2.5. INVARIANCE OF PHASE DIAGRAMS OF OLIGOMER BLENDS

Figure 2.10 shows phase diagrams of the blends comprising oligobutadieneurethane diacrylate (OBUDA) and oligomers of the homologous series of n-hydroxyethylene dimethacrylates (n-HEDA). The diagrams are plotted in two forms, using (i) the conventional T versus w coordinates (curves 2–4) and (ii) the ΔT versus w coordinates, where $\Delta T = \text{UCST} - T$ (curve 1). As can be seen, in the latter case the binodals of all the systems under consideration overlap to yield a single curve. This implies that the boundaries of the domains of existence of stable solutions for these mixtures would coincide, provided that the binodals are shifted by the values equal to differences between their critical solution temperatures. It was found [86] that all thermodynamic functions of mixing also acquire equal values at a fixed distance ΔT from UCST.

As an example, Fig. 2.11 shows the temperature variation of the interaction parameter χ and its enthalpy (χ_H) and entropy (χ_S) components calculated for the above-mentioned blend systems and plotted in the χ versus ΔT coordinates. As is seen, the curves coincide to within the experimental error (characteristic of the method of compatibility determination employed in [85]) and the uncertainty of differentiation, which is inavoidable in the calculation of χ values. It was shown [85] that a similar pattern is observed for many other systems comprising a polymer and a homologous series of oligomers, reported in the literature, if the corresponding phase diagrams are plotted in the χ versus ΔT coordinates and the thermodynamic parameters are reduced to the ΔT value.

These results suggest that the phase diagrams and thermodynamic functions of mixing are invariant with respect to UCST for the blend systems with solvents represented by homologous series of oligomers.

This circumstance is of importance from theoretical standpoint, primarily because it provides a quantitative criterion for assessing

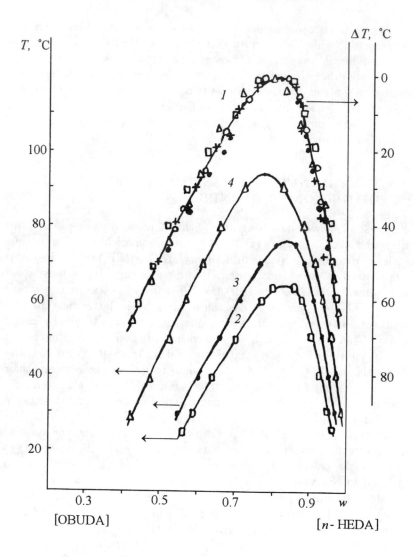

Figure 2.10. Phase diagrams of the OBUDA–n-HEDA blends for $n = 2$ (*2*), 3 (*3*), and 4 (*4*); curve 1 represents a generalized phase diagram plotted in the ΔT versus w coordinates.

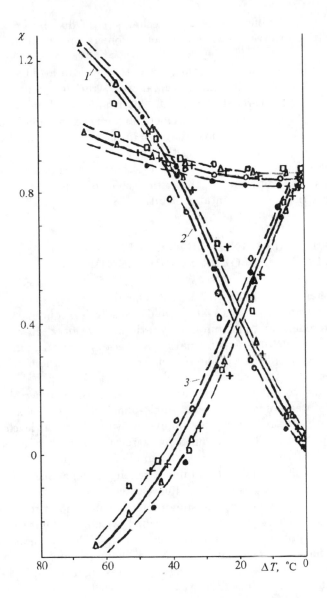

Figure 2.11. The plots of (1) χ_{12}, (2) χ_H, and (3) χ_S versus ΔT for the OBUDA–n-HEDA blends with $n = 2$ (o), 3 (•), 4 ($\triangle$), and 13 (+).

the generality and difference of the mechanisms of dissolution and phase formation involved in the mixing of oligomers with homologous solvents.

The practical significance of this fact is that we can avoid a considerable amount of routine experimental work, rather complicated and ambiguously interpreted, devoted to determination of phase diagrams for all mixtures of a given polymer (oligomer) and homologous solvents. Now, once a single phase diagram in the series is known, any other is obtained by merely shifting the known binodal by a distance equal to the difference of UCST values for the known and unknown systems. Note that determining the critical solution temperature encounters no serious experimental difficulties [68].

2.6. STRUCTURAL INHOMOGENEITY OF SINGLE-PHASE OLIGOMER BLENDS

Once the phase diagram of a system is available, we can determine the number of phases in this system, and the solution concentration at a given temperature and component ratio in any phase. However, phase diagrams provide no information concerning the structure of individual phases and the morphology of the system as a whole, which data are necessary for technological calculations.

Let us first consider the structure of single-phase oligomer blend systems. As was noted above, the single-phase character of a system does not necessarily imply its homogeneity on the other levels of structural hierarchy. The supermolecular structure of both solution I and solution II (see Fig. 2.1) is rather complicated. One of the main factors responsible for the system inhomogeneity is the appearance of supermolecular formations having a fluctuational nature.

In the existing literature, the supermolecular inhomogeneities occurring in polymeric systems are described within the framework of various, sometimes rather ambiguous notions. These notions involve both the terms borrowed from classical chemistry and the newly formed concepts, such as associates, agglomerates, aggregates, blobes, domains, clusters, cybotaxises, etc. Of course, each of these has the right to exist, provided that its physical meaning is clearly defined.* In this monograph, the supermolecular structures formed in liquid

---

* Unfortunately, the same terms are sometimes differently treated by various authors. Moreover, in some cases one term is given different meanings in various parts of the same text.(see, e.g., [51]).

oligomer blend systems are mostly described using two terms: "associate" and "cybotaxis". In order to define these items, we will employ the common and distinguishing features as described in [73]. The common features are as follows: (a) these terms refer to supermolecular structures in liquid oligomer blend systems, which are formed by long-living fluctuations; (b) the lifetimes of these fluctuations exceed the mean statistical values; (c) both items do not represent phase formations. The distinctions are as follows: (a) an associate can be composed of either like or unlike molecules, while a cybotaxis may comprise only like molecules; (b) cybotaxises always exhibit anisotropy in the spatial arrangement of molecules, while associates can be isotropic. Note that the reasons that lead to the formation of associates and cybotaxises in a system (van der Waals forces, dipole–dipole interactions, hydrogen bonds, "tails" in the distribution of random fluctuations, metastable and critical states of the system, etc.) may be either the same of different.

The question as to whether a spatial ordering and a short-range order may exist in oligomeric liquids arose in 1960s because certain anomalies were observed for the initial curing rates of oligoester acrylates [23]. However, it was not until recent years that sufficiently reliable experimental grounds were obtained which confirmed the existence of "labile precursors with kinetically favorable or unfavorable order" in oligomeric systems [24, 25, 73, 83].

Spontaneous formation of the structural elements of short-range order is an "anti-entropy" process that seems to be in contradiction with the laws of thermodynamics, according to which an equilibrium system must tend to maximum entropy, that is, to the maximum disorder. However, this is only an apparent contradiction, because the minimum free energy of an oligomer blend system is attained under conditions that are far from complete disorder. This notion was experimentally and theoretically justified (see the preceding section) and considered in detail elsewhere [71–73, 78–82].

There are several models of inhomogeneous structure of liquids. One of these is a cybotaxic model developed by Frenkel in 1920s [26]. In recent years, the model has been successfully applied in materials science [26], in particular to polymeric materials [25]. It would be expedient to refine the main points of this model.

Cybotaxises ($\chi\iota\beta\omega\tau o\sigma$, a Greek word meaning ark, was introduced into science by Stewart at the beginning of this century) are microscopic domains of liquid in which molecules exhibit a spatial order of certain type, different from the structure of surrounding liquid. Much later, a close concept of "cluster" (meaning bunch, swarm) was developed; this term is relative to cybotaxis, but does not imply any spatial orientation (anisotropy). The presence of preferred orientation of molecules in a cybotaxis is compensated by their random arrangement in the neighboring microvolumes. The cybotaxic regions are instable by themselves, spontaneously appearing at one point, then disappearing, and arising again at another site. The lifetime of a cybotaxis depends on the composition and temperature and is determined by the energy of intermolecular physical bonds. It is suggested that an ordered state in cybotaxises exists for a longer time as compared to the lifetime of ordinary fluctuations—random deviations of the concentration, density, etc., from mean statistical values. In this respect, the cybotaxises are much closer to associates, complexes, and solvates—the concepts employed for the description of solutions in the physical chemistry—differing from these by the fluctuational nature. The thermal motion of molecules results in that cybotaxises have no clearly defined boundaries and no physical interfaces. Therefore, we cannot consider them as the phase formations. Moreover, decomposition of a cybotaxis implies no strictly determined parameters of state, thus differing from the classical phase transition.

Thus, the cybotaxic model reflects a microinhomogeneous structure of liquid within the framework of a single-phase state. This is a point of principal importance, in particular, because analyses of spinodal and nucleation mechanisms of the phase transitions in oligomeric systems [28, 29, 70] may involve a confusion of terms. Moreover, the concept of microinhomogeneity is sometimes unconsciously replaced by microheterogeneity (in the sense of phase separation).

The number of molecules in a cybotaxis can vary. Accordingly, the dimensions and lifetimes of cybotaxises may exhibit a scatter. An attempt to evaluate these parameters was made in [21]. The results of optical and electron-microscopic examinations of oligoester acrylates and their mixtures, together with calculations of maximum length and cross-sections of oligomer molecules, showed that cybotaxises must exhibit a distribution with respect to probable dimensions and the packing density of molecules. It was found that the diameter of these fluctuations may vary from tens to a few hundreds of Engstrums, and the number of molecules in such a local microinhomogeneity may vary from 6 to 40. These estimates agree well with the data reported in [28], where the X-ray diffraction study of liquid oligostyrenes showed evidence of the presence of ordered domains with a periodicity of 17 Å. Estimates obtained for the cybotaxis lifetime τ in oligomeric liquids [83] indicated that, within certain temperature interval, we have $\tau = 10^{-3}$–10^{-5} s.

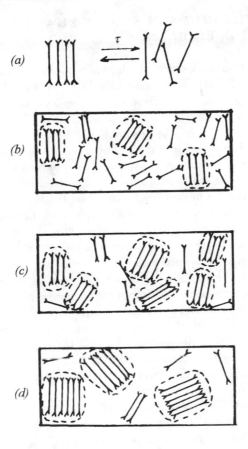

Figure 2.12. Schematic diagram illustrating a cybotaxic model of oligomer blends. Bars represent oligomer molecules, τ denotes the cybotaxis lifetime. The cybotaxic liquids may differ by the cybotaxis lifetime (a), by the number of cybotaxises per unit volume (b versus c), and by the dimensions of cybotaxises (b versus d).

Various cybotaxic structures occurring in oligomer blend systems are schematically depicted in Fig. 2.12. For simplicity, molecules of the second component (i.e., of the medium in which the cybotaxis forms and "lives") are omitted. The diagram is quite illustrative and requires no detailed comments. It should be noted that this scheme has proved to be very fruitful for the analysis of equilibria in single-phase oligomer blend systems.

2.7. EQUILIBRIUM IN SINGLE-PHASE OLIGOMER BLENDS

The relaxation properties of oligomer–polymer systems were studied by pulsed NMR techniques [30, 31]. The experiments were performed on films of *cis*-polyisoprene rubber swelled in tetramethylene dimethacrylate. In the literature, samples of this type are usually referred to as the "equilibrium-swelled" films. However, experimental data showed that the procedure of equilibrium swelling did not ensure that the sample attained the state of thermodynamic equilibrium.

Figure 2.13 shows variation of the characteristic time of transverse nuclear relaxation of protons T_2 and their fraction P with the storage (exposure) time τ_{exp} for the blend samples and the initial components. Note that the experimental curve representing decay of the transverse magnetization for the initial (individual) oligomer (Fig. 2.13, curve *1*) can be described by exponential law with a single time constant T_2^{oligo} independent of τ_{exp}. The decay of magnetization in the initial rubber is described by a superposition of two exponential functions with the corresponding transverse relaxation times, $(T_2^{rub})'$ and $(T_2^{rub})''$, that are also independent of the exposure time (lines *2* and *3*).

After swelling, the samples exhibit three component relaxation times T_2. One of these is attributed to relaxation of the oligomer molecules, and the two others are assigned to the rubber. Note that the relaxation time of oligomer molecules in the blend $(T_2^{oligo})_{blend}$ is lower than that in the initial oligomer, because mobility of the oligomer molecules is lower in the presence of rubber component. The relaxation times of rubber in the mixture, $(T_2^{rub})'_{blend}$ and $(T_2^{rub})''_{blend}$, are higher as compared to that in the initial unswelled rubber (an increase in the mobility of rubber molecules is due to plasticization). This behavior is quite natural. An unexpected result was that the T_2 value of each component in the blend became dependent on τ_{exp} (as noted above, the characteristic times of nuclear relaxation before mixing were independent of the exposure). Initially, the $(T_2^{oligo})_{blend}$ values increase 3–4 times with increasing exposure and then (in approximately 5 days) return to the starting level. The $(T_2^{rub})''_{blend}$ value also grows initially, then exhibits a low-pronounced maximum and attains a limiting level after a 10-day exposure. During this, the fraction of protons P corresponding to each characteristic time also gradually varies, which is a further evidence for a redistribution of protons between different ensembles.

Similar results in this respect were obtained from investigations of the relaxation characteristics of other polymer–oligomer and oligo-

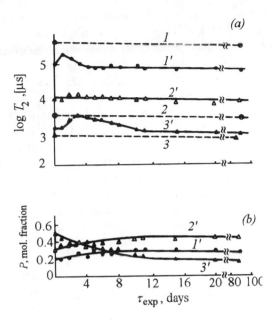

Figure 2.13. The plots of $\log T_2$ (a) and P (b) versus the exposure duration τ_{exp} for *cis*-polyisoprene swelled in tetramethylene dimethacrylate ($1'$–$3'$) and the initial components (1–3). See the text for explanations.

mer–oligomer systems [30, 31, 71, 72]. Long-term processes of attaining equilibria in oligomer blend systems were also observed in the study of curing kinetics in the systems [32, 33, 73].

The whole body of data published in the literature suggests that storage of a swelled system, even after attaining a constant weight, is accompanied by structural rearrangements. This conclusion implies, in turn, that the process of attaining thermodynamic equilibrium in oligomer blend systems has a stepwise character, comprising at least two stages.

In the first stage observed after bringing the initial components into contact, the mutual diffusion leads to leveling of their concentrations over the macroscopic volume of the system. In the case under consideration, the process is limited by the time of swelling to a constant weight (for our system, this required about 8 h). In the general case, the establishing of equilibrium in the first stage is determined by the time to complete mutual dissolution of the components in the coexisting phases.* A driving force for this process consists in minimization of the gradient of chemical potential in the system, occurring in the course of mixing (swelling, dissolution, etc.).

After the constant component concentrations are reached in the coexisting phases, the second stage begins in which another type of equilibrium is established in the system, whereby the supermolecular formations exhibit leveling with respect to their dimensions and lifetimes, and their number approaches the average value. This is a slow process (10–12 days for the systems under consideration) involving no further changes in the chemical potential. The latter circumstance is related to the fact that the cybotaxic rearrangement (which is also referred to as a modification of the fluctuation-associative organization of local inhomogeneities in the oligomer blend systems) is caused by a change in the ratio of combinatory and non-combinatory entropy components involving, within the accuracy of experimental determinations, no change in the total entropy of the system [71].

The above results and their conclusions are important from both theoretical and practical standpoint. Once the multistage character of attaining equilibrium in oligomer blend systems is known, a process engineer may exclude or at least restrict the factors (time, temperature, viscosity) controlling the establishing of thermodynamic equilibrium in a given composition, thus increasing the reproducibility of conditions and the quality of products.

2.8. EQUILIBRIUM AND MORPHOLOGY CONTROL IN TWO-PHASE OLIGOMER BLENDS

The attaining of equilibrium in a heterogeneous oligomer blend system has a much more complicated character as compared to that in a single-phase blend. Besides all the features known in homogeneous systems, which naturally take place in each one of the coexisting phases in the heterogeneous system, there are additional processes such as the interphase mass exchange by diffusion and the phase separation (layer formation).

The interphase diffusion in oligomer blend systems is virtually not studied, although it is this process that controls the coalescence

---

* It should be emphasized that not only oligomer molecules diffuse into the polymer matrix, but the rubber molecules pass into the liquid phase as well.

of drops in emulsions. Apparently, the known mechanisms of the interphase transport of polymer diffusants, which were described in [19, 35],are also valid in oligomer systems.

As was mentioned above (see Section 2.1), the rate of phase separation processes in a heterogeneous system is a function of many variables. Because the phase separation is a process of importance in practice (directly related to the properties of materials), we will analyze this in more detail. For simplicity, the analysis will refer to a two-phase blend.

In the course of mixing of the components of an oligomer blend system, in which the component concentrations correspond to a point situated "below the spinodal" (see Fig. 2.1), a two-component blend having the form of emulsion is always formed. In this blend, the concentrations of components in the corresponding phases are determined only by the temperature. It should be recalled that the equilibrium values of w' and w'' at $T = $ const correspond to the points located on the left and right branches of the binodal, respectively. The morphology (i.e., the size and distribution of drops) of the emulsion, formed upon mixing of a particular pair of components in the given oligomer blend, depends not only on the nature of components (eventually determining the surface tension of solutions I and II) and their ratio, but as well on the method of blending (forced mixing, mutual dissolution, etc.) the type of blending equipment (rollers, plate stirrers, paint grinders, etc.), and the regime of blending (duration, rate, and temperature of mixing or the mutual solvent removal). In other words, the degree of dispersion of a two-phase oligomer blend systems determined by the energy "pumped" into the system in the course of mixing. Restricted experimental data available in the literature [21, 35–37, 72] indicate a broad particle size distribution in such systems, ranging from 0.01 to 20–30 μm.* Provided all other conditions are equal, the blend morphology and the nature of the dispersion medium depend on the concentration of oligomer. This is clearly illustrated by Fig. 2.14. As the oligomer concentration in the blend increases, the ratio of volume fractions of solutions II and I varies, and the dimensions of disperse inclusions tend to grow. There are certain values $c = c_i$ at which a phase inversion takes place, whereby a phase that was disperse at $c < c_i$ becomes continuous at $c > c_i$.**

---

* Note that each phase in a two-phase oligomer blend system, that is, each drop in the emulsion and in the emulsion medium, exhibits inhomogeneity, in particular, cybotaxises of much smaller dimensions.

** Conditions (c, T) can be selected at which both phases will be continuous [10, 12].

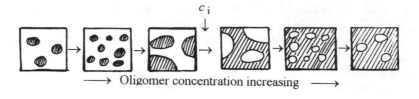

$$\longrightarrow \text{ Oligomer concentration increasing } \longrightarrow$$

▨ - Solution of polymer in oligomer (II)

☐ - Solution of polymer in polymer (I)

Figure 2.14. Variation of the morphology of a two-phase blend system with increasing oligomer concentration (c_i represents the point of phase nversion).

Oligomer blend systems may differ by their morphologies developed by the moment of termination of the blending process, Moreover, the morphology may further vary during the subsequent storage of the blend (see Fig. 2.15). The temporal stability of the morphology is determined primarily by viscosity of the dispersion medium. If the viscosities of solutions I and II are close to one another, the question as to which phase is continuous is insignificant. On the contrary, if the viscosities of the coexisting phases are differing (as noted above, these may differ by several orders of magnitude in the oligomer blend systems), the knowledge of what is the dispersion medium is of primary importance for a process engineer. Indeed, if the medium is represented by a low-viscosity phase (e.g., by solution II), the dispersed particles exhibit rapid coalescence leading to a complete phase separation. The time to complete phase separation estimated for such oligomer blend systems [21, 38] ranges from 2–10 min to 1–2 days, which is comparable with the duration of technological operations. Therefore, in this case we cannot ignore the rate of phase separation determining the current parameters of system morphology.

A different situation is observed when the dispersion medium is represented by a highly viscous phase, such as a solution of oligomer in polymer (solution I). In this case, the morphology developed in the stage of mixing may remain unchanged during an infinitely long period of time. Experimental data were reported, which showed that

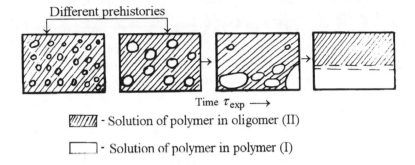

Different prehistories

Time τ_{exp} $\longrightarrow$

▨▨▨ - Solution of polymer in oligomer (II)

☐ - Solution of polymer in polymer (I)

Figure 2.15. Schematic diagram illustrating phase separation in polymer–oligomer systems for $\eta_1 \gg \eta_2$. See the text for explanations.

some "raw" rubber–oligoester acrylate systems exhibited virtually no changes in the colloidal-disperse structure upon a 5-year storage (at room temperature) [21, 35], while PVA–oligomer blends were stable for at least one year [39]. However, this quasi-equilibrium is disturbed if some of the kinetic restrictions are removed (e.g., the temperature is increased). Then the viscosity of dispersion medium begins to decrease and, at a certain temperature, reaches a level at which it is no more controlling the mobility of particles of the dispersion phase. As a result, the system looses stability, the drops of emulsion begins to coalesce, and the situation follows that for a low-viscosity dispersion medium considered above.

Interesting possibilities to control the morphology of oligomer blend systems are offered by using surfactants as emulsifying agents. A highly valuable experience, which can be also useful in oligomer blends systems, was gained in studying latexes [40] and some other disperse systems [15, 41]. However, there are special features inherent only in reactive oligomer blends. For example, it was demonstrated [42, 43] that rubber–oligoester acrylate systems can be modified by nonionogenic surfactants represented by monomethacrylic esters of unsaturated alcohols with the general formula:

$$CH_2 = \overset{\overset{\displaystyle CH_3}{\displaystyle |}}{C} - \overset{\overset{\displaystyle O}{\displaystyle \|}}{C} - O - (CH_2)_n - CH_3$$

These are not only capable to emulsify the system, but may eventually participate in the process of blend curing by copolymerization with oligoester acrylate. This may exclude undesirable side effects in the cured blends [14], which are typically introduced by usual surfactants [44]. As for the morphology control in oligomer blend systems, it was experimentally shown that small (1–5%) additives of the reactive surfactants can improve the compatibility of polymer–oligomer mixtures.

Figure 2.16 shows the concentration dependence of the optical density of blends between *cis*-polyisoprene and α-trimethacryl-ω-methacryl-pentaerythritol (dimethacryl- pentaerythritol adipinate) in the presence and absence of a 2% surfactant additive. As is seen, adding surfactant leads to an approximately three-fold increase in the critical oligomer concentration (c_{cr}) at which the system is considered [10, 45] as passing from single- to two-phase state. Evidently, these small surfactant additives cannot affect the thermodynamic compatibility of blends. This was confirmed by an experiment, in which spontaneous dissolution of the given oligoester acrylate in *cis*-polyisoprene was found to vary only at a surfactant concentration above 10%. An increase in the transparency of film samples of the oligomer blends, and the change of c_{cr} was due to a decrease in the emulsion particle size in the presence of surfactant, rather than due to a variation in the chemical affinity of the blend components. Because the nephelometric technique is capable of detecting only particles with dimensions exceeding 1000 Å, all smaller phase formations existing in the oligomer blend system remain unobserved by this method.

The above example is of interest not only by clearly illustrating one possible error that can be involved in the analysis of experimental data on the compatibility of oligomer blends. In addition, this case demonstrates original possibilities to control the blend structure on the colloid-disperse level by rather simple technological methods.

2.9. RELATIONSHIP BETWEEN THE STRUCTURE AND TECHNOLOGICAL PROPERTIES OF OLIGOMER BLENDS

In this Section we will analyze correlations between the structural organization of oligomer blend systems on various hierarchical levels and the macroscopic characteristics realized in a system before curing (i.e., those determining to a considerable extent the technological properties of the system).*

---

* Technological properties (workability) of oligomer blends depend on

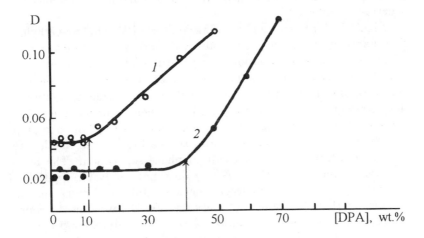

Figure 2.16. Concentration dependence of the optical density of blends between *cis*-polyisoprene and α-trimethacryl-ω-methacryl-pentaerythrite (dimethacryl-pentaerythrite adipinate, DPA) (*1*) in the presence and (*2*) in the absence of surfactant.

2.9.1. Translational Diffusion

The laws established by studying diffusion in oligomer–oligomer and polymer–oligomer systems [19, 21, 34, 39, 46] form a general notion about the character of mass transfer in oligomer blends. We can distinguish three correlations from the whole body of data, which are of special importance from the technological standpoint: (i) a jumplike change in the values of mutual diffusion coefficients in the region of phase transitions; (ii) a change in the diffusion rate accompanying the passage of a system from glassy to rubberlike state; and (iii) depen-

---

many factors [47], including viscosity, deformation rate, adhesion to materials of equipment, coefficients of friction, thermal conductivity, heat capacity, diffusion, density, etc. Various criteria have been formulated that determine the workability on the basis of combinations of the above parameters. A conventional approach consists in using a similarity theory to determine a "dominating" property for a given material in the process considered.

dence of the rate of diffusion on the molecular and supermolecular structure of diffusant.

Figure 2.17 presents a phase diagram for a PVC–hydroxyethylene dimethacrylate blend in comparison with the mutual diffusion coefficients D_v measured at various temperatures and component ratios [21]. A significant feature is the existence of break points on the concentration dependences of D_v at the values of φ_1 (volume fraction of oligomer) equal to the solubility limit of oligomer in polymer. This jumplike change in the character of diffusion is due to a boundary, separating the regions of solutions I and II (see Fig. 2.1) in the domain of mixing, which forms as a result of phase separation. The break observed in the curves $D_v = f_1(\varphi_1)$ does not imply termination of the exchange of diffusing molecules between different phases, but only indicates that the gradient of mass transfer is altered in the course of diffusion as a result of variation of the thermodynamic properties of the system. This phenomenon is naturally explained within the framework of the phenomenological theory of diffusion [19], according to which the rate of mass transfer is a function of the chemical potential (μ_1) and is determined for a binary system by the formula

$$D_v = D_s(\varphi_1/RT)(\partial\mu_1/\partial\varphi_1), \qquad (2.5)$$

where D_s is the coefficient of self-diffusion. Because the phase separation implies that $\partial\mu_1/\partial\varphi_1 \to 0$, the D_v value also decreases at the binodal.

It is important to note that, in contrast to the coefficient D_s, the coefficient of self-diffusion at $T > \text{UCST}$ in the system under consideration increases with φ_1 [39]. This observation is of certain value for the technological practice. Indeed, it implies that equilibrium (temperature-controlled at a constant φ_1) in the system is achieved mainly by diffusion of the oligomer from polymer-depleted to polymer-rich phase (i.e. from solution II to solution I).

The presence of a maximum in the region of solutions ($\varphi_1 < \varphi_{cr}$) on the plot of D_v versus φ_1 is explained by competition of two factors. On the one hand, by increase in the mobility of the matrix in the course of oligomer dissolution (as a result of plasticization). This is accompanied by a decrease in the "resistance" of the diffusion medium and an increase in the diffusion rate. On the other hand, spontaneous mixing of the components increases thermodynamic nonideality of the solution, which is connected with the formation of associates and cybotaxises of the diffusing molecules. This naturally leads to a decrease in the velocity of translational diffusion.

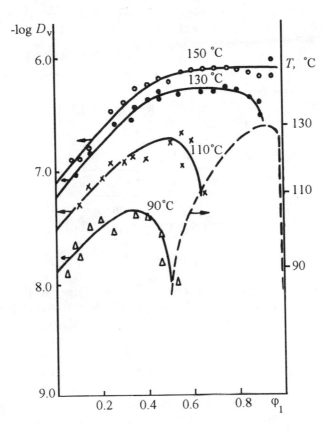

Figure 2.17. Phase diagram (dashed curve) for a PVC–HEDA blend system and the composition dependence of the mutual diffusion coefficients D_v measured at various temperatures (indicated at the curves) [21].

Another feature of the diffusion in oligomer blend systems consists in a jumplike variation of the temperature parameter of diffusion E_{act} (effective activation energy) upon the passage of the system from glassy to rubberlike state. The value of E_{act} calculated for $\varphi_1 \to 0$ in the temperature region below T_g is two times the value obtained at temperatures above T_g [48]. Dependence of the diffusion velocity on the molecular mass of oligomer in the temperature region far from

Table 2.2. The values of parameters K and b in equation (2.6) for various temperatures and conpositions of a PVC–oligo(esteracrylate) system

Parameter	T, °C	φ_1			
		0	0.1	0.2	0.3
K	150	$4.0 \cdot 10^{-3}$	$5.0 \cdot 10^{-3}$	$9.5 \cdot 10^{-3}$	$1.8 \cdot 10^{-2}$
	130	$1.6 \cdot 10^{-3}$	$3.3 \cdot 10^{-3}$	$5.2 \cdot 10^{-3}$	$1.1 \cdot 10^{-2}$
	110	$7.9 \cdot 10^{-4}$	$1.6 \cdot 10^{-3}$	$2.5 \cdot 10^{-3}$	$4.2 \cdot 10^{-3}$
	90	$3.2 \cdot 10^{-4}$	$6.3 \cdot 10^{-4}$	$1.2 \cdot 10^{-3}$	$1.3 \cdot 10^{-3}$
b	90–150	2.1 ± 0.4	1.9 ± 0.2	2.1 ± 0.1	2.2 ± 0.2

the binodal is satisfactorily described by the relation

$$D_v = KM^{-b}, \tag{2.6}$$

where b depends on the geometry (size and shape) of the diffusant molecules and K is determined by the local microviscosity of the medium (retardation parameter).

As is seen from Table 2.2, the value of b is independent of the temperature and composition of the oligomer blend system. Nor it depends on the length of the oligomer block [48]. At the same time, the value of K grows with increasing temperature and φ_1.

Thus, the character of the translational diffusion of components in oligomer blends is determined by the following factors:

(i) By the phase state of the system. In single-phase oligomer blends, the diffusion rate exhibits a nonmonotonic variation which is determined by competition of the thermodynamic and relaxation processes acting in the opposite directions and varying in the course of diffusion. In the region of the transition from homogeneous to heterogeneous state of the system, the rate of mutual diffusion exhibits a sharp drop due to a decrease of the chemical potential near the binodal.

(ii) By the physical state of the system. The rate of diffusion in elastomers is higher by 1–2 orders of magnitude, and the diffusion activation energy is lower 1.5–2 times, as compared to the values in glassy matrices. During the transfer of oligomers into glassy polymers, the latter may acquire elasticity, and the diffusion velocity can sharply increase in the course of the process.

(iii) By the molecular structure of diffusant. Dependence of the diffusion rate on the molecular mass of oligomer is described by a power function. Numerical values of the coefficients required for the calculation of D_v were determined for typical oligomer systems [19, 21, 29, 39, 74].

2.9.2. Rotational Diffusion

In oligomer blend systems, the rotational diffusion was studied in much detail by methods of dipole polarization and pulsed NMR [21, 31, 39, 49, 91]. The results of these investigations revealed several correlations between the parameters of electric and magnetic relaxation and the phase structure of the system.

Figure 2.18 shows the concentration dependence of the relaxation time τ of the dipole polarization, the activation energy ΔU of this process, the temperature T_m at which the dielectric losses exhibit a peak, and the values of loss tangent $\tan \delta_m$ at this temperature reported for a cis-polyisoprene–trihydroxyethylene dimethacrylate system [49].* The arrow indicates a critical oligomer content w_{cr} (in weight fractions) corresponding to the phase separation at a temperature of the sample preparation (20°C).

As is seen, all parameters of rotational diffusion exhibit a jump-like variation in the vicinity of w_{cr}. However, the main reasons for this behavior of various parameters (as well as that of the parameters of translational diffusion considered above) can be different. The increase of $\tan \delta_m$ and the decrease of $\log \tau$ at $w_1 > w_{cr}$ are due to increasing proportion of the parcels of a polymer-rich phase (in which the constants of rotational mobility of the molecules are higher than those in solution I). The decrease of $\log \tau$ and T_m with increasing concentration of the oligomer after transition of the system into a two-phase state (at $w_1 > w_{cr}$) reflects a decrease in the retarding effect of the second component on the rotational diffusion of oligomer molecules, which is caused by a decrease in the volume fraction of the polymer-rich phase.

It should be emphasized that the parameters of rotational diffusion, as determined by the method of relaxation of the dipole polarization, are integral characteristics (averaged over the total number of molecules in both phases). Therefore, the laws revealed by this method cannot reflect the character of variation of the mobility of individual molecules occurring in various phases. This information was gained by the pulsed NMR technique.

---

* Direct experimental data presented in [49] confirm the segregation of oligomer molecules and the appearance of inhomogeneities in single-phase oligomer systems.

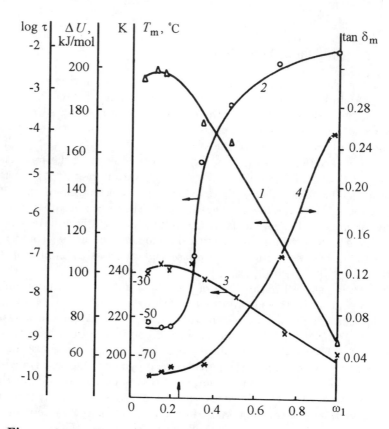

Figure 2.18. The plots of (*1*) relaxation time $\log \tau$, (*2*) activation energy ΔU, (*3*) dielectric loss peak temperature T_m, and (*4*) $\tan \delta_m$ versus the total oligomer content w_1 for a *cis*-polyisoprene–trihydroxyethylene dimethacrylate blend system. The relaxation measurements were performed at a frequency of 1 kHz [49].

Figure 2.19 shows the concentration dependence of the spin-spin relaxation time T_2 in a PVC–trihydroxyethylene dimethacrylate system [38]. For the oligomer contents below 5%, the pulsed NMR technique yields a single relaxation time for the given oligomer blend, $T_2 = 25$–30 ms, which is 1.5–2 times the value for a pure (unmixed) PVC, and five orders of magnitude lower as compared to the value for pure oligoester acrylate. The fact that only one transverse relaxation time is detected at $w_1 < 5\%$ in the system containing molecules of two kinds with different nuclear relaxation times is indicative of the restricted possibilities of this experimental method. For the oligomer content above 5% and up to $w_1 = w_{cr}$ (that is, in the region of single-phase states), the decay of magnetization is described by two T_2 values: a "long-term" relaxation time T_{2a} characterizing the rotational modes of oligomer molecules, and a "short-term" relaxation time T_{2b} describing the rotational mobility of macromolecules.* The fact that the T_{2a} value is much smaller than T_2 of the pure oligomer can be explained by retarded mobility of the oligomer molecules dissolved in polymer, which may be due to segregation of these molecules.

In the region of two-phase states of the system ($w_1 > w_{cr}$) the curve of the transverse magnetization decay displayed four characteristic relaxation times: two "short-term" values (T_{2b}' and T_{2b}'') and two "long-term" values (T_{2a}' and T_{2a}'') characterizing the mobility of oligomer molecules (subscript 'a') and polymer macromolecules (subscript 'b') in the phases of solution I (single prime) and solution II (double prime). The T_{2b}'' value in the oligomer-rich phase is about ten times higher than T_{2b}', which is explained by low concentration of the polymer molecules (see the phase diagram in Fig. 2.13) and, hence, a low proportion of the polymer–polymer contacts. At the same time, a low number of the polymer–polymer contacts (also due to a small polymer concentration) in the phase of solution II accounts for an increase of the mobility of oligomer molecules in this phase as compared to solution I. Nevertheless, polymer plays a significant role in restricting the mobility of oligomer molecules: T_{2a}'' is about ten times smaller than T_2 of the pure oligomer.

A very important conclusion following from these experimental data is that the T_{2a}', T_{2a}'', T_{2b}' and T_{2b}'' values are constant for each particular phase. This result gives unambiguous evidence that the component concentrations in coexisting phases are independent of the total oligomer content at $w_1 > w_{cr}$, thus confirming the equilibrium state of the system.

We must also mention another feature of the oligomer blend system, which was observed in the study of dipole relaxation [49]. These experimental data showed evidence that small additives ($< 3\%$) of oligomers may significantly affect the supermolecular structure of the second component, which forms upon blending. It was found that, in the presence of small amounts of some oligoester acrylates, cis-polyisoprene exhibits retarded segmental mobility both in the glassy

---

* The "short" and "long" relaxation times are usual terms of a "laboratory" slang employed in the analysis of NMR data.

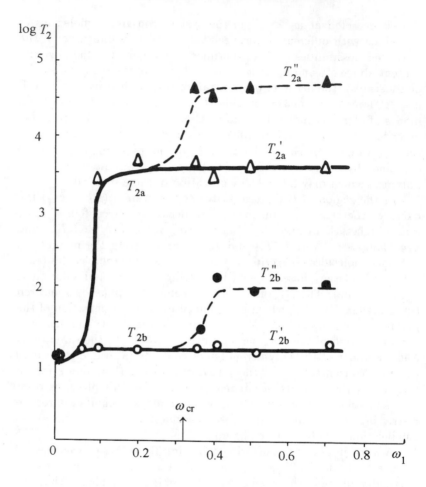

Figure 2.19. The plot of relaxation time T_2 versus the total PVC content for various components of a cis-polyisoprene–trihydroxyethylene dimethacrylate blend system at 40°C: (o) PVC; (Δ) 3-HEDA; open and black symbols refer to phases I and II, respectively.

state (below the α-transition) and in the rubberlike state (above T_g). Whatever are the reasons for the effect, this behavior reflects an increase in the local structural inhomogeneity of the system, leading eventually to a change in the macroscopic properties.

2.9.3. Adsorption on Solid Surfaces

This phenomenon determines to a significant extent the adhesion characteristics of the blend. Therefore, the effect must be controlled within wide limits, from complete suppression to maximum enhancement. Sometimes, a process engineer has to solve mutually exclusive tasks. For example the adsorption must be reduced in the stage of component mixing in order to decrease (eliminate) sticking of the oligomer blend to walls of the stirring and molding equipment. At the same time, the adsorption ability must be retained on a level sufficiently high to provide for the cohesion strength of the blend (shape stability), the "substrate-blend" adhesion strength (for the oligomer blends used in the glue compounds), or ensure a strong interphase interaction (for the blends used as binders in filled materials or reinforced composites).

Modern notions about the mechanisms of adsorption of polymers and oligomers are well formulated in reviews [50–52]. The existing theories of the phase separation in polymer solutions and blends in the presence of solid surfaces, which have been extensively developed in recent years, are summarized in [75–77]. The features of adsorption in the blend systems, in particular, in the oligomer blends, include (i) selectivity, (ii) dependence on the solution concentration, and (iii) difference in the sorption kinetics of the blend components. All these properties are mostly determined by the aggregative mechanism of sorption.

An essential feature of the aggregative mechanism of sorption consists in that not only (and not mostly) individual molecules, but their aggregates as well, pass to the surface of adsorbent. These aggregates are represented by associates and cybotaxises. As a result, a time-dependent equilibrium is established between the aggregated and nonaggregated molecules. The aggregation characteristics (concentration, dimensions, and lifetimes of the aggregates) depend on the solution concentration (the oligomer concentration differs significantly for the solutions I and II), the phase organization of the blend (a much higher organization level in the metastable and critical states), and the thermodynamic affinity of components (a higher degree of aggregation for the components with low affinity). Therefore, the knowledge of these relationships would allow us not only to explain rather involved correlations observed in oligomer blend systems, but to predict their manifestations as well.

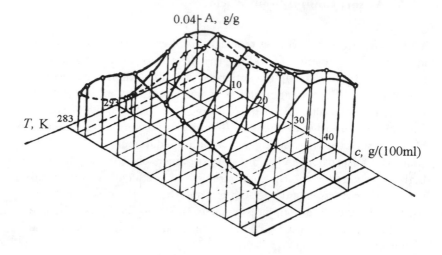

Figure 2.20. Isotherms of adsorption of an ED-5 epoxy oligomer from toluene on glass spheres [51].

Figures 2.20–2.22 present data on the dimensions of aggregates, aggregation constants,* and isotherms of adsorption on glass spheres obtained for an oligomer–monomer mixture (epoxy oligomer ED-5 in toluene) at various temperatures and oligomer concentrations [51]. A comparative analysis of these data reveals a clear correlation between the value of adsorption and the character of structure formation in the system. Variations in the size of aggregates and the extent of interactions between them, depending on the temperature and composition of the mixture, determine the shape of the isotherm and the complicated temperature and concentration dependences of the oligomer concentration in the system.

Figure 2.23 shows the kinetics of oligoethylene glycol adsorption from acetone on aerosil and soot [51]. The magnitudes of extrema on the kinetic curves grow with increasing oligomer concentration in

---

* The aggregation constant K_a was calculated as the ratio K'/K in the equilibrium system $n \leftrightarrow Kn_a + K'n_f$, where n is the total number of molecules per unit volume, n_a is the number of molecules in aggregates, and n_f is number of free molecules [51].

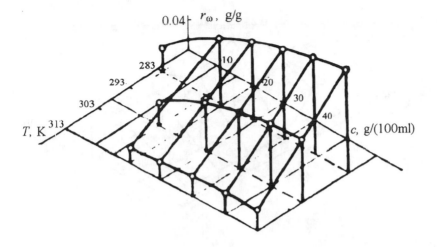

Figure 2.21. Average aggregate size r_w for an ED-5 epoxy oligomer adsorbed at various temperatures and concentrations from
toluene on glass spheres [51].

the solution: the higher the concentration, the greater is the size of
aggregates primarily adsorbed. The following decrease in the value
of adsorption is caused by the establishing of adsorption equilibrium
between aggregates of various dimensions, and is also related to the
kinetics of conformational changes of the molecules adsorbed on the
particle surface.

2.9.4. Viscosity

This property was thoroughly studied in oligomer blend systems of
many types [10,21, 53–58, 88–90]. It was established that the shear
viscosity of oligomer blends is determined primarily by the phase organization of the mixture. Naturally, the laws of viscous flow in solutions
are principally different from those in the disperse systems.

   In the region of solutions, oligomer blends of various types
deformed in various regimes exhibit a variety of manifestations of
their rheological properties. For example, experiments showed both
Newtonian and non -Newtonian character of flow, depending on the
temperature and shear velocity. Investigations of the viscosity as a

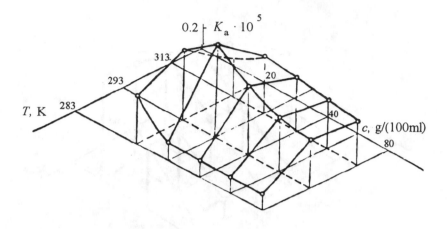

Figure 2.22. Aggregation constants of an epoxy oligomer ED-5 adsorbed at various temperatures and concentrations from toluene on glass spheres [51].

function of the ratio of oligomer blend components sometimes yielded additive curves, but more frequently revealed deviations from additivity, which could be either positive or negative. In other words, single-phase oligomer blend systems exhibited the same spectrum of flow patterns which was known for highly viscous polymeric liquids [10, 16, 53].

The laws of viscosity variation are usually explained on a qualitative level within the framework of the Frenkel's kinetic theory of liquids [26] using the entanglement network [57] or cybotaxic structure [88] models. Application of the scaling hypothesis [59] to analysis of the dynamic properties of oligomer blend solutions allowed the concept of entanglements to be presented in the form of quantitative relations [60]. The cybotaxic model proved to be extremely fruitful for description of the laws of recently discovered temperature hysteresis in the viscous flow of oligomers [89, 90].

Figure 2.24 shows the plots of viscosity versus temperature for oligobutadieneurethane diacrylate (OBUDA) with polydisperse molecular mass distribution, measured in two regimes: heating (curve *1*) and cooling (curve *2*). As is seen, the viscosity of OBUDA samples measured at the same temperature is higher in the cooling regime

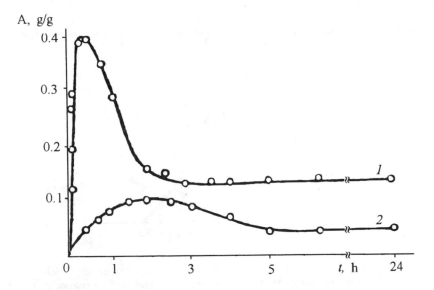

Figure 2.23. Kinetics of oligo(ethylene glycol) adsorption on (*1*) aerosil and (*2*) soot from acetone solutions with the oligomer concentrations (*1*) 3.84 g/(100 ml) and (*2*) 0.76 g/(100 ml) [51].

than in the heating mode. Thus, the viscosity of the blend samples that were initially heated to temperatures above the test point. is higher than that of the samples not pre-exposed at higher temperatures. This phenomenon is explained by the activation character of the formation of equilibrium cybotaxic structure [89], whereby certain energy must be first spent to destroy a structure randomly developed in the preceding stage, before passing to a thermodynamically allowed supermolecular structure possessing a higher viscosity.

If the further investigations would confirm that this experimental observation is generally valid for the other systems, and if the activation nature of the formation of supermolecular structure in single-phase oligomer blends would be proved, many of the data published in the literature will have to be critically reassessed. This refers to the values of rheological constants calculated with no allowance for the possible thermodynamically nonequilibrium state of oligomer and polymer solutions during the measurements.

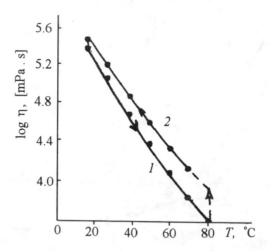

Figure 2.24. The plots of viscosity versus temperature for OBU–DA samples measured in the heating (curve *1*) and cooling (curve *2*) modes at $\gamma = 54$ s^{-1}.

The phenomenon of viscosity hysteresis is not less important from the standpoint of practice. For example, a process engineer must take into account that overheating of a blend in the course of technological operations preceding the stage of molding may alter (even not leading to any undesired chemical transformations) the supermolecular structure, which would increase the viscosity of a liquid phase and detrimentally affect its workability. Therefore, the temperature regime of oligomer blend treatment must be controlled within the region bounded by the lower curve in Fig. 2.24.

In the region of two-phase systems, the viscosity of oligomer blends obeys the laws of flow known for dispersions and emulsions, and depends on the number, size, and shape of dispersed inclusions [10, 15, 53–55].

In the region of phase transitions, experiments revealed a nonmonotonic variation of viscosity. In some cases, the viscosity of blends exhibited a sharp drop, whereas in the other cases, it showed a jumplike increase. The drop of viscosity upon the phase transition appears as a quite natural result in view of the above-mentioned macroscopic properties of oligomer blends. On the contrary, an increase in viscosity cannot be directly explained within the framework of these notions. For example, Fig. 2.25 shows a concentration dependence of the effective viscosity for a *cis*-polyisoprene–*β*-HEDA blend system [62]. In the region of small total oligomer contents (far from the binodal), the viscosity exhibits a sharp initial raise and then decreases to the additive values. The relative increment of viscosity depends on the molecular characteristics of blend components, the temperature, the shear rate and stress, the method of testing, the sample prehistory, etc. [21, 58]. This anomalous phenomenon (unusual features include, in particular, a growth of the blend viscosity with increasing content of the less viscous component) is also related to the phase and supermolecular organization of the oligomer blend. However, the relationship is indirect [63], rather than explicit as in the case of decreasing viscosity or changing diffusion and adsorption characteristics of the blends.

The physical concept that allows us to explain the experimental anomalies is essentially as follows. As is known, long-living fluctuations of the concentration spontaneously appear in a system occurring in the critical or metastable state [60]. If this "critical" system is subjected to deformation (as it is in determining the viscosity by the capillary or rotational techniques), the shear energy is partly spent to destroy (diffusion smearing) the fluctuational supermolecular structures, whereas in the other states of the system, the energy is entirely consumed in the events of molecular transfer in the direction of applied stress. In other words, a greater force is required to ensure the flow event in a system occurring in the critical or metastable state than in a usual or the so-called regular solution,* because a part of the applied energy is spent "in vain" (from the standpoint of viscosity variation). Note that the structurized systems feature, in addition to these processes, the osmotic suction forces that facilitate the transport of molecules [64].

Therefore, the viscosity of oligomer blend systems near the phase separation boundary is determined as a sum of the viscosity of the regular solution plus positive corrections for the critical and metastable state and a negative osmotic correction. The real situation is additionally complicated by the fact that components of an oligomer blend system are usually distributed with respect to their molecular masses. This, in turn, implies that real oligomer blends always contain molecules belonging to various parts of the phase diagram [61]. Thus, a certain proportion of molecules may occur in the critical state and

---

* That is, in the solution occurring far from a critical point and a metastable region, which is assumed to have no longwave fluctuations.

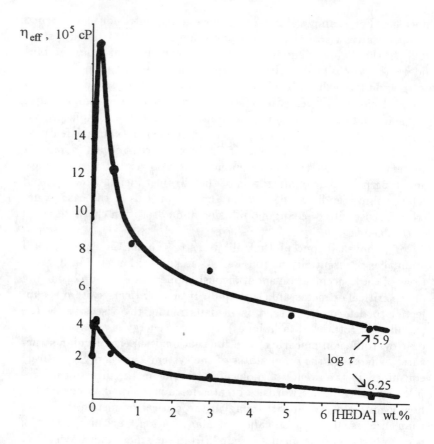

Figure 2.25. The plot of the effective viscosity versus total oligomer content for a *cis*-polyisoprene–*3*-HEDA blend system at various shear stress levels.

participate in the formation of large-scale fluctuations, whereas another part will occur in a metastable state and form other types of fluctuations, the third part will be dispersed in the regular solution, etc.

  In the simplest case, the macroscopic viscosity of an oligomer blend system composed of polydisperse components can be expressed as follows:

$$\eta_b = \eta_{reg} + \Delta\eta_{cr}\varphi_{cr} + \Delta\eta_m\varphi_m - \Delta\eta_{os}\varphi_{os}, \tag{2.7}$$

where η_{reg} is the viscosity of the regular solution; $\Delta\eta_{cr}$ is the critical correction describing an increase in the viscosity of a part φ_{cr} of the molecules, occurring in the critical region; $\Delta\eta_m$ is the metastable correction for a part φ_m of the molecules occurring near the spinodal; and $\Delta\eta_{os}$ is the correction for the osmotic forces acting upon the fraction Δ_{os}, of the molecules.

Evidently, the terms $\Delta\eta_{cr}$ and $\Delta\eta_m$ control an increase in the viscosity of the oligomer blend system, while $\Delta\eta_{os}$ produces a decrease in the total viscosity. Competition between the positive and negative contributions determines the viscosity of the blend η_b. Quantitative relations for the calculation of corrections were derived by Manevich [64, 66, 67]. Each expression is a complicated power function of the shear stress and velocity, concentrations, molecular masses, and molecular mass distribution parameters of the blend components, temperature, and some other parameters determining the phase and supermolecular organization of the blend. Even very small variations of these characteristics lead to changes in the macroscopic viscosity measured in experiment. These circumstances account for the scatter of data reported by various authors for the same oligomer blend systems.

For the oligomer blend processing technology, it is important that knowledge of the laws of flow allows the viscosity to be varied within wide limits by changing the process conditions.

An apparently intriguing result that is also worth mentioning was observed in the study of PVC–oligoester acrylate blends [68]. Figure 2.26 shows a rheological kinetic curve constructed by the data of viscosity measurements performed in the scanning mode (heating rate, 1 K/min). As is seen, the plot of viscosity as a function of the temperature exhibits a maximum, although all the known laws of polymer rheology suggest that the viscosity must only decrease with increasing temperature in the absence of chemical interactions. However, a thorough analysis reveals that the experimental plot shows nothing extraordinary, even assuming no chemical interactions in the system. Indeed, experiments performed in [68] were devoted to the study of viscosity changes in the course of component mixing (emulsion-polymerized granulated PVC was blended with 3-HEDA). Thus, the data refer to a nonequilibrium system. In the first stage of blending (swelling of polymer granules in oligomer), we observe the flow in a disperse system. Because the rate of the oligomer diffusion into polymer grows with the temperature, the degree of swelling and, hence, the size of dispersed particles, increase. Consequently, the viscosity increases as well in accordance with the well-known Einstein's relation.

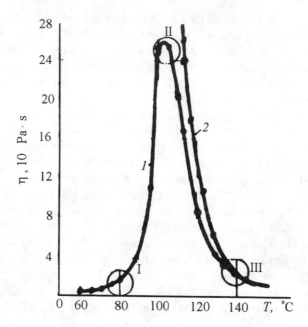

Figure 2.26.   Rheological kinetic curves of the PVC–*3*-HEDA
(60 : 40) blend measured in the heating (1) and cooling (2) regime
for the rate of temperature variation 1 K/min [68].

After the formation of a plastisol monolith, that is, after completion of
the dissolution and the formation of equilibrium single-phase system,
the viscosity decreases with increasing temperature—now in complete
agreement with the laws of flow in solution. Measurements in the reverse (cooling) regime showed only increasing viscosity, because the
system remained single-phase in the entire temperature interval studied.

The above examples show that only the knowledge of phase diagrams and the kinetics of attaining the equilibrium state would allow a
process engineer to select a *priori* the optimum regime of the oligomer
blend processing.

REFERENCES

1. *Entsiklopediya polimerov* (Encyclopedia of Polymers), Moscow: Sov. Entsiklopediya, 1977, vol. 3, p. 433 (in Russian).
2. Manson, J.A. and Sperling, L.H., *Polymer Blends and Composites,* New York: Plenum, 1976.
3. Tager, A.A., *Fizikokhimiya polimerov* (Physical Chemistry of Polymers), Moscow: Khimiya, 1978 (in Russian).
4. Paul, D.R., in: *Polymer Blends,* Paul, D.R. and Newman, S., Eds., New York: Academic, 1978.
5. Flory, P., *Statistical Mechanics of Chain Molecules,* New York: Wiley, 1969.
6. Macknight, W.J., Karasz, F.E., and Fried, J.R., in: *Polymer Blends,* Paul, D.R. and Newman, S., Eds., New York: Academic, 1978.
7. Krause, S., in: *Polymer Blends,* Paul, D.R. and Newman, S., Eds., New York: Academic, 1978.
8. Sanchez, I.C., in: *Polymer Blends,* Paul, D.R. and Newman, S., Eds., New York: Academic, 1978.
9. Papkov, S.P., *Ravnovesie Faz v Sistemakh Polimer–Rastvoritel'* (Phase Equilibria in Polymer–Solvent Systems), Moscow: Khimiya, 1981 (in Russian).
10. Kuleznev, V.N., *Smesi Polimerov* (Polymer Blends), Moscow: Khimiya, 1980 (in Russian).
11. Nesterov, A.E. and Lipatov, Yu.S., *Obrashchennaya Gazovaya Khromatografiya v Termodinamike Polimerov* (Reversed-Phase Gas Chromatography in Thermodynamics of Polymers), Kiev: Naukova Dumka, 1976 (in Russian).
12. Lebedev, E.V., in: *Fizikokhimiya Mnogokomponentnykh Polimernykh Sistem* (Physical Chemistry of Multicomponent Polymeric Systems), Kiev: Naukova Dumka, 1986, vol. 2, p. 5 (in Russian).
13. Kwei, T.K., Nishi, T., Roberts, R.F., *Macromolecules,* 1974, vol. 7, no. 5, p. 667.
14. Rebinder, P.A., *Fiziko-Khimicheskaya Mekhanika* (Physico-Chemical Mechanics), Moscow: Nauka, 1979 (in Russian).

15. Lipatov, Yu.S., *Kolloidnaya Khimiya Polimerov* (Colloid Chemistry of Polymers), Kiev: Naukova Dumka, 1984 (in Russian).
16. Malkin, A.Ya. and Chalykh, A.E., *Diffuziya i Vyazkost' Polimerov. Metody Izmerenuya* (Diffusion and Viscosity of Polymers: Methods of Measurement), Moscow: Khimiya, 1979 (in Russian).
17. Rakovskii, A.V., *Vvedenie v Fizicheskuyu Khimiyu* (Introduction to Physical Chemistry), Moscow: OITL, 1938 (in Russian).
18. Nesterov, A.E. and Lipatov, Yu.S., *Fazovoe Sostoyanie Rastvorov i Smesei Polimerov. Spravochnik* (Phase State of Polymer Solutions and Blends. A Handbook), Kiev: Naukova Dumka, 1987 (in Russian).
19. Chalykh, A.E., *Diffuziya v Polimernykh Sistemakh* (Diffusion in Polymer Systems) Moscow: Khimiya, 1987 (in Russian).
20. Chalykh, A.E. *et al., Dokl. Akad. Nauk SSSR,* 1978, vol. 238, no. 4, p. 893.
21. Mezhikovskii, S.M., Structure and Properties of Polymer–Oligomer Systems and Related Composites, *Doctoral (Tech. Sci.) Dissertation,* Moscow, 1983 (in Russian).
22. Kotova, *et al., Vysokomol. Soedin., Ser. A,* 1982, vol. 24, no. 3, p. 460.
23. Berlin, A.A., Kefeli, T.Ya., and Korolev, G.V., *Poliefirakrilaty* (Polyester Acrylates), Moscow: Nauka, 1967 (in Russian).
24. Berlin, A.A., Korolev, G.V., Kefeli, T.Ya., and Sivergin, Yu.M., *Akrilovye Oligomery i Materialy na Ikh Osnove* (Acrylic Oligomers and Related Materials), Moscow: Khimiya, 1983 (in Russian).
25. Mezhikovskii, S.M., *Vysokomol. Soedin., Ser. A,* 1986, vol. 29, no. 8, p. 1571.
26. Frenkel', Ya.I., *Kineticheskaya Teoriya Zhidkosti* (Kinetic Theory of Liquids), Leningrad: Nauka, 1975 (in Russian).
27. Baum, B.A., *Metallicheskie Zhidkosti* (Melal Liquids), Moscow: Nauka, 1973 (in Russian).
28. Shilov, V.V. and Lipatov, Yu.S., in: *Fizikokhimiya Mnogokomponentnykh Polimernykh Sistem* (Physical Chemistry of Multicomponent Polymeric Systems), Kiev: Naukova Dumka, 1986, vol. 2, p. 5 (in Russian).
29. Rozenberg, B.A., *Problemy Fazoobrazovaniya v Oligomer–Oligomernykh Sistemakh* (Problems of Phase Formation in Oligomer–Oligomer Systems), Chernogolovka: Akad. Nauk SSSR, 1986 (in Russian).
30. Mezhikovskii, S.M., *et al., Dokl. Akad. Nauk SSSR,* 1983, vol. 302, no. 4, p. 878.
31. Lantsov, V.M., Structural-Kinetic Inhomogeneity of Molecules in Oligomeric and Polymer–Oligomer Systems, and Related Cross-Linked Polymer Networks, *Doctoral (Chem. Sci.) Dissertation,* Moscow, 1989 (in Russian).
32. Mezhikovskii, S.M., *Nekotorye Problemy Fizikokhimii Polimer–Oligomernykh Sistem i Kompozitov na Ikh Osnove* (Selected Problems of Physical Chemistry of Polymer–Oligomer Systems and Related Composites), Chernogolovka: Akad. Nauk SSSR, 1986 (in Russian).
33. Mezhikovskii, S.M., Chalykh, A.E., and Zhil'tsova, L.A., *Vysokomol. Soedin., Ser. B,* 1986, vol. 28, no. 1, p. 53.

34. Maklakov, A.I., Skirda, V.D., and Fatkulin, N.F., *Samodiffuziya v Rast-vorakh i Rasplavakh Polimerov* (Self-Diffusion in Polymer Melts and Solutions), Kazan: Kaz. Gos. Univ., 1987 (in Russian).
35. Mal'chevskaya, T.D., Formation and Properties of Vulcanizates Based on Rubber–Oligomer Compositions, *Cand. Sci. (Chem.) Dissertation*, Moscow, 1980 (in Russian).
36. Rebrov, A.V., Features of Structure Formation in Rubber–Oligo(ester-acrylate) Systems, *Cand. Sci. (Chem.) Dissertation*, Moscow, 1983 (in Russian).
37. Lipatov, Yu.S., Mezhikovskii, S.M., and Shilov, V.V., *Kompozit. Polim. Mater.*, 1986, no. 28, p. 31.
38. Mezhikovskii, S.M., *Polimer–Oligomernye Kompozity* (Polymer–Oligo-mer Composites) Moscow: Znanie, 1989.
39. Kotova, A.V., Phase Structure of Poly(vinyl chloride)–Oligo(esteracry-late) Systems and Related Composites: Thermodynamic and Kinet-ic Laws of Structure Formation, *Cand. Sci. (Chem.) Dissertation*, Moscow, 1988 (in Russian).
40. *Novye Sinteticheskie Lateksy i Teoreticheskie Osnovy Lateksnoi Tekhno-logii* (New Synthetic Latexes and Theoretical Backgrounds of Latex Technologies), Chernaya, V.V., Ed., Moscow: TsNIITENefteKhim, 1973.
41. Tolstaya, S.N. and Shabanova, S.A., Primenenie *Poverkhnostno-Aktiv-nykh Veshchestv v Lakokrasochnoi Promyshlennosti* (Application of Sur-factants in Lacquer-Paint Industry), Moscow: Khimiya, 1976 (in Rus-sian).
42. Panchenko, V.I., Effect of Oligo(esteracrylates) on the Rubber Pro-cessing and the Properties of Related Rubbers, *Cand. Sci. (Tech.) Dissertation*, Volgograd, 1974 (in Russian).
43. Frenkel', R.Sh. and Panchenko, V.I., *Modifikatsiya Rezin Oligoefirakri-latami* (Modification of Elastomers by Oligo(esteracrylates)), Moscow: TsNIITENefteKhim, 1981.
44. Abramzon, A.A., *Poverkhnostno-Aktivnye Veshchestva* (Surfactants), Leningrad: Khimiya, 1976 (in Russian).
45. Rabek, J., *Experimental Methods in Polymer Chemistry*, Chichester (U.K.): Wiley, 1980.
46. Hopfenberg, H.B. and Paul, D.R., in: *Polymer Blends*, Paul, D. and Newman, S., Eds., New York: Academic, 1979.
47. Vostroknutov, *et. al.*, *Pererabotka Kauchukov i Rezinovykh Smesei* (Processing of Rubber and Elastomer Mixtures), Moscow: Khimiya, 1980 (in Russian).
48. Kotova, A.V., Chalykh, A.E., and Mezhikovskii, S.M., *Vysokomol. Soedin., Ser. A*, 1983, vol. 25, no. 1, p. 163.
49. Gladchenko, S.V., Study of the Dipole Relaxation and Structure of Polymer–Oligomer Compositions, *Cand. Sci. (Phys.-Math.) Disserta-tion*, Leningrad, 1983 (in Russian).
50. Fleer, G.J. and Lyklema, J., in: *Adsorption From Solutions on the Sol-id/Liquid Interface*, Parfitt, G.D. and Rochester, C.H., Eds., London: Academic, 1983.

51. Todosiichuk, G.T. and Lipatov, Yu.S., in: *Fizikokhimiya Mnogokomponentnykh Polimernykh Sistem* (Physical Chemistry of Multicomponent Polymeric Systems), Kiev: Naukova Dumka, 1986, vol. 1, p. 130 (in Russian).
52. Semenovich, G.M. and Lipatov, Yu.S, in: *Fizikokhimiya Mnogokomponentnykh Polimernykh Sistem* (Physical Chemistry of Multicomponent Polymeric Systems), Kiev: Naukova Dumka, 1986, vol. 1, p. 186 (in Russian).
53. Vinogradov, G.V. and Malkin, A.Ya., *Reologiya Polimerov* (Rheology of Polymers), Moscow: Khimiya, 1980 (in Russian).
54. Van Oene, H., in: *Polymer Blends,* Paul, D.R. and Newman, S., Eds., New York: Academic, 1978.
55. Lipatov, Yu.S, in: *Dokl. 1 Vses. Konf. po Khimii i Fiziko-Khimii Polimerizatsionnosposobnykh Oligomerov* (Proc. of the First All-Union Conf. on the Chemistry and Physical Chemistry of Polymerizable Oligomers), Chernogolovka: Akad. Nauk SSSR, 1977, vol. 1, p. 59 (in Russian)
56. Shumskii, V.F., in: *Fizikokhimiya Mnogokomponentnykh Polimernykh Sistem* (Physical Chemistry of Multicomponent Polymeric Systems), Kiev: Naukova Dumka, 1986, vol. 2, p. 279 (in Russian).
57. Malkin, A.Ya. and Kulichikhin, S.G., *Reologiya v Protsessakh Obrazovaniya i Prevrashcheniya Polimerov* (Rheology of the Processes of Formation and Transformation of Polymers), Moscow: Khimiya, 1985 (in Russian).
58. Mezhikovskii, S.M., Vasil'chenko, E.I., and Shaginyan, Sh.A., *Usp. Khim.,* 1986, no. 11, p. 1867.
59. De Gennes, P., *Scaling Concepts in Polymer Physics,* Ithaca (USA): Cornell University, 1979.
60. Grosberg, A.Yu. and Khokhlov, A.R., *Statisticheksys Fizika Makromolekul* (Statistical Physics of Macomolecules), Moscow: Nauka, 1990 (in Russian).
61. Kaningsveld, R., *Adv. Colloid. Interface Sci.,* 1968, vol. 2, p. 151.
62. Mezhikovskii, S.M. *et al., Dokl. Akad. Nauk SSSR,* 1976, vol. 229, no. 2, p. 410.
63. Mezhikovskii, S.M., Polym. News, 1991, vol. 16, no. 3, p. 91.
64. Mitlin, V.S. and Manevich, L.I., Vysokomol. Soedin., Ser. B, 1985, vol. 27, no. 6, p. 409.
65. Mezhikovskii, S.M., in: Polimery-90 (Polymers-90), Chernogolovka: Akad. Nauk SSSR, 1991, vol. 1, p. 174.
66. Manevich, L.I., Shaginyan, Sh.A., and Mezhikovskii, S.M.,Dokl. Akad. Nauk SSSR, 1981, vol. 258, no. 1, p. 142.
67. Manevich, L.I., Mitlin, V.S., and Shaginyan, Sh.A., Khim. Fiz., 1984, no. 2, p. 283.
68. Yaroshevskii, S.A., Polymer–Oligomer Composites Based on Linear Polymers and Oligo(ester acrylates), Cand. Sci. (Tech.) Dissertation, Moscow, 1986 (in Russian).
69. Gibbs, J.W., Selected Thermodynamic Works, Moscow: Goskhimizdat, 1950 (in Russian).

70. Manevich, L.I. and Shaginyan, Sh.A., Spinodal'nyi Raspad Binarnykh Smesei Oligomerov v usloviyakh Khimicheskoi Reaktsii (Spinodal Degradation of Binary Oligomer Mixtures under Conditions of Chemical Reactions), Chernogolovka: Ross. Akad. Nauk., 1994 (in Russian).
71. Mezhikovskii, S.M., in: *Papers Presented at the Intern. Rubber Conf. (IRC-94)*, Moscow: 1994, vol. 3, p. 83.
72. 72. Erbich, Yu.R., in: *Papers Presented at the Intern. Rubber Conf. (IRC-94)*, Moscow: 1994, vol. 3, p. 195.
73. Mezhikovskii, S.M., *Kinetika i Termodinamika Protsessov Samoorganizatsii v Oligomernykh Smesevykh Sistemakh* (Kinetics and thermodynamics of Self-Organization Processes in Oligomer Blend Systems), Chernogolovka: Ross. Akad. Nauk., 1994 (in Russian).
74. Avdeev, N.N., Diffusion and Phase Equilibrium in Oligomeric and Polymer–Oligomer Systems, *Cand. Sci. (Phys.-Math.) Dissertation*, Moscow: Inst. Fiz. Khim. Ross. Akad. Nauk, 1983 (in Russian).
75. De Gennes, P., *Rev. Mod. Phys.*, 1985, vol. 57, p. 827.
76. Sullivan, D.E. and Telo da Gamma, M., in: *Fluid Interfacial Phenomena*, Croxton, C., Ed., New York: Wiley, 1986, p. 45.
77. Dolinnyi, A.i. and Ogarev, V.A., *Usp. Khim.*, 1986, vol. 57, no. 9, p. 1769.
78. Polak, L.S. and Mikhailov, A.S., *Samoorganizatsiya v Neravnovesnykh Fiziko-Khimicheskikh Sredakh* (Self-Organization in Non-Equilibrium Physico-Chemical Media), Moscow: Nauka, 1990 (in Russian).
79. Nicolis, G. and Prigogine, I., *Self-Organization in Nonequilibrium Systems*, New York: Wiley, 1977.
80. Furukawa, H., *Adv. Phys.*, 1985, vol. 34, no. 6, p. 703.
81. Hoshimoto, T., *Phase Transitions*, 1988, vol. 12, no. 1, p. 47.
82. Binder, K., in: *Materials Science and Tehnology*, Weinheim, H.P., Ed., New York, 1991, vol. 5, p. 405.
83. Korolev, G.V. and Berezin, M.P., in: *Tez. Plen. Stend. Dokl. V Konf. po Khimii i Fizikokhimii Oligomerov* (Abstracts of Plenary and Poster Reports Presented at the 5th Conf. on the Chemistry and Physical Chemistry of Oligomers), Chernogolovka: Ross. Akad. Nauk., 1994, p. 13 (in Russian).
84. Prigogine, I., *The Molecular Theory of Solutions*, Amsterdam: North-Holland, 1957.
85. Mezhikovskii, S.M., Kotova, A.V., and Repina, T.B., *Dokl. Ross. Akad. Nauk*, 1993, vol. 333, no. 2, p. 197.
86. Repina, T.B., Kotova, A.V., Tseitlin, G.M., and Mezhikovskii, S.M., *Vysokomol. Soedin., Ser. B*, 1995, vol. 35, no. 9, p. 1557.
87. Tinius, C., Plasticizers, ???
88. Vasil'chenko, E.I. and Mezhikovskii, S.M., *Vysokomol. Soedin., Ser. A*, 1989, vol. 31, no. 7, p. 1362.
89. Mezhikovskii, S.M. and Vasil'chenko, E.I., *Dokl. Ross. Akad. Nauk*, 1994, vol. 339, no. 5, p. 627.
90. Vasil'chenko, E.I., Repina, T.I., and Mezhikovskii, S.M., in: *Tez. Plen. Stend. Dokl. V Konf. po Khimii i Fizikokhimii Oligomerov* (Abstracts of Plenary and Poster Reports Presented at the 5th Conf. on the Chemistry and Physical Chemistry of Oligomers), Chernogolovka: Ross. Akad. Nauk., 1994, p. 13 (in Russian).

91. Nazarova, I.I., Kulagina, T.P., Andrianova, Z.S., and Baturin, S.M., *Issledovanie Dinamiki i Struktury Lineinykh i Sshitykh Poliuretanov po YaMR-Spadam Namagnichennosti* (Studying the Dynamics and Structure of Linear and Cross-Linked Polyurethanes by the NMR Magnetization Decay Technique), Chernogolovka: Ross. Akad. Nauk., 1991 (in Russian).

3 KINETICS OF POLYMERIZATION REACTIONS AND MECHANISM OF STRUCTURE FORMATION DURING THE CURE OF OLIGOMER BLENDS

3.1. CURING KINETICS

Curing of an oligomer blend involves chemical conversion of a liquid reactive system to a solid material. Only two variants are possible when formulating a system on the basis of commercially important oligomer blends: either (a) only one of the components (the oligomer) of the formulation is reactive during the molding or (b) both components are reactive.

In the first case (the second component is inert), curing of monomer–oligomer or oligomer–oligomer blend leads to a system in which the resultant polymer may be dissolved, swollen, plasticized, or dispersed in a chemically inert solvent (either a monomer or an oligomer).* As a result of curing of polymer–oligomer systems (in these systems the polymer is a chemically inert component), one of the variants of the well-known temporary plasticization principle [1, 2] takes

---

* These situations are similar to polymerization with precipitation, dispersion polymerization, preparation of plastisols, etc. In addition, technologically, these systems may be considered as offering the only way to introduce plasticizers into densely cross-linked polymer networks or chemically inert ingredients into rigid linear matrices (by analogy with the preparation of oil-filled rubbers).

place, and the polymerization of oligomer in a polymer matrix leads to a polymer–polymer blend, in which the newly formed cross-linked polymer is dispersed in a matrix of the initial linear polymer.

In the second case, when both components of an oligomer blend are reactive, a polymer–polymer blend is always eventually formed. The structures of such blends may be very different. Indeed, a blend may comprise a mixture of linear polymers, cross-linked or a linear and a cross-linked polymers, a mixture of block or graft copolymers, etc.. However, in all cases, the resultant polymer blend is heterogeneous.

When both components of a blend are reactive, (i) the functional groups of both components may be chemically identical or, at least, may react according to the same mechanism or (ii) the functional groups of each of the components may be different and, therefore, the curing of these components proceeds according to different mechanisms.

In the first variant, which is usually implemented with the aim of improving the properties of the material based on a particular oligomer, a monomer or an oligomer is added that is different from the major component in its nature, the length of oligomer chain, or the number of functional groups. For example, to improve the impact strength of the glasses based on trioxyethylene dimethacrylate (TGM-3), the formulated compound is doped with tridecaoxyethylene dimethacrylate (TGM-13) or bis(trioxyethylene)phthalate/dimethacrylate (MDF-2). Three-dimensional copolymerization of the doped compound leads to a network, in which the functionality of all network junctions is the same, whereas some cross-linkages are longer than in the network obtained by the homopolymerization of TGM-3.* As a result, the modified network is more flexible and soft.

Another example is provided by the synthesis of polyurethanes by migration polymerization of diisocyanates with glycols using oligoglycols of different molecular masses. As a result, block copolyurethanes are obtained, which, depending on the lengths of oligoclycol-derived blocks, are characterized by different tendencies to crystallization [3]. When bifunctional monomers (glycols, diamines, hydrazine, and others) are also added into such formulations in addition to oligoglycols, segmented polyurethanes that possess the properties of thermoplastic elastomers are obtained.

For various classes of oligomer blends, the examples of imparting valuable properties to materials by using mixtures of oligomeric com-

---

* The real structure of the resultant networks is much more complex. We will elaborate on this problem in the discussion below.

ponents, whose curing mechanisms are similar, are numerous. However, oligomer blends comprising the components that are cured according to different mechanisms are of significantly greater applied and scientific interest. These oligomer blends are used to prepare such important materials as impact-resistant plastics, hybrid binders, simultaneous IPNs, rubber–oligomer elastomeric systems, cellular plastics, varnishes, etc. These materials have been discussed in the modern literature in detail (e.g., see the monographs [4–8, 18]).

It is essential that during the cure of this type of oligomer blends, the chain propagation rates for each component of the blend are different. Therefore, the polymer (linear or cross-linked) that is formed first performs as a matrix, within which the polyreactions involving the second component take place. Because the difference in the propagation rate constants is usually rather significant, the polymer matrix forms almost immediately, that is, from the very start the process may be treated as the curing of a polymer–oligomer system. The matrix significantly affects the curing kinetics of oligomers. Indeed, the ratio between the rate constants of the elementary events corresponding to chain propagation, chain termination, and chain transfer changes. The conditions for phase separation also change. Moreover, as a result of the variation of the molecular characteristics of the components during the curing and the concomitant variation of thermodynamic parameters of the blend, additional type of phase separation, viz., separation at a level involving macromolecules of alien components, comes into play.

These factors make the curing kinetics of oligomer blends significantly different from the classical regularities that describe polymerization and polycondensation [9–11]. Comparison with the curing kinetics for individual oligomers, which are characterized by inherent specific features [12, 13], also reveals that the curing of oligomer blends follows a different pattern.

Figures 3.1–3.3 illustrate some particular cases relevant to the curing of oligomer blends. As can be seen, the rate of polymerization depends on the prehistory of the oligomer blend; the curing kinetics follows a stepwise pattern; and the plot of conversion in polymerization versus the curing temperature shows a minimum [14–16]. These examples may be supplemented by quite a number of other illustrations of the features specific to the curing kinetics of oligomer blends. Some of those features are inherent to distinct systems and are related to the peculiarities of the chemical structure of the components, whereas the other features pertain to general regularities characteristic of oligomer blends.

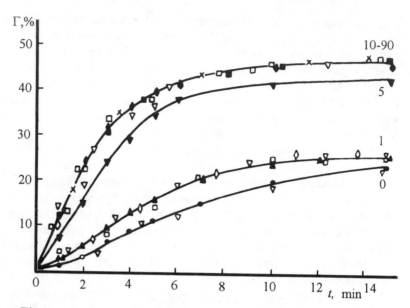

Figure 3.1. Kinetics of the polymerization of tetramethylene dimethacrylate in a matrix of SKI-3 isoprene rubber for the blends that were stored at different temperatures (figures at the curves denote storage time in days).The samples were obtained by equilibrium (36.5%) swelling at 20°C. The curing temperature T_{cure} was 107°C. The initiator was AIBN (0.5%).

These general regularities involve (1) the relationship between the curing kinetics and non-equilibrium character of the initial state; (2) the relationship between the curing kinetics and phase organization of the initial oligomer blend; (3) the relationship between the curing kinetics and the specific features of supermolecular organization of liquid oligomer blends; (4) the relationship between the kinetics of polyreactions and specific features of phase separation, which inevitably accompanies the curing.

In spite of the fact that the curing of pure individual oligomers has been thoroughly discussed in the literature [12, 13], we considered it necessary to recall here the major regularities that control the formation of polymer structure in the homopolymerization of oligomers. This will provide a logical sequence in describing the kinetics and mechanism of the curing of oligomer blends and will be helpful in revealing the corresponding specific features.

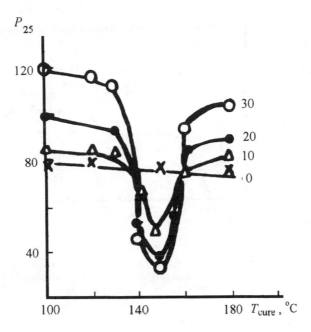

Figure 3.2. Penetration depth at $25°C$ (P_{25}) as a function of T_{cure} for the blend of trioxyethylene dimethacrylate with BND-60/90 bitumen with different added amounts of the oligomer (figures at the curves correspond to percentage of the oligomer). The initiator was dicumyl peroxide (2%). Curing time, 2 hours.

3.2. CURING OF REACTIVE OLIGOMERS

In fact, curing of oligomers with the aim of obtaining linear polymers has not found practical application except for a few reactions (the most important is urethane formation) which were discussed in [3, 17]. Processes that lead to formation of networks are much more important.

Formation of a network suggests that the functionality of the reactive groups of oligomers is greater than two. An alternative way to prepare a network is to react a bifunctional oligomer with a curing agent that contains more than three reactive groups. The curing reactions may involve polymerization, polycondensation, or polyaddition. The reactions may be initiated either by heat, light, radiation, or by

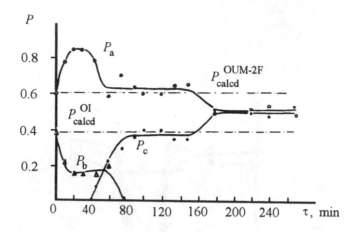

Figure 3.3. Variation of the fractions of the protons (P_a, P_b, and P_c) during the course of curing of the blend of oligoimide (OI) with oligourethane methacrylate (OUM-2F) at 125°C.

material initiators and catalysts. Anyway, formation of a network is a complex chemical process, which is highly specific for various classes of oligomers; this case was described in detail in the corresponding monographs [1, 3, 8, 12, 13, 17, 19]. As a rule, formation of a network is accompanied by side reactions and phase and state transitions. The fine mechanisms of polyreactions have not been yet established for various classes of reactive oligomers. However, general regularities that are independent of the mechanism of the process have been identified.

The kinetics of polyreactions involving oligomers is usually described in terms of several common stages: initiation (or any other means of activation), chain propagation, and chain termination. Chain transfer reactions may also take place. As an example, Fig. 3.4 illustrates two particular cases of network formation. Case (a) describes the initiation and chain propagation in the three-dimensional polymerization of a tetrafunctional unsaturated oligomer.* Case (b) illustrates the

---

* We speak of oligomers with two terminal double bonds. Scission of each unsaturated bond gives rise to two chain propagation sites. Therefore, one bifunctional molecule of oligoester acrylate, oligomaleinate/fumarate or

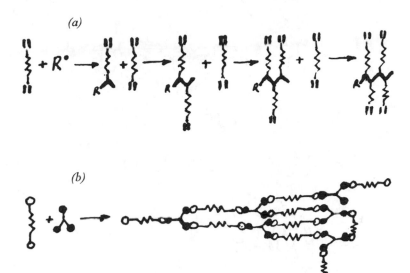

Figure 3.4. Schematic illustration of the formation of a polymer network. (a) Polymerization of a tetrafunctional unsaturated oligomer initiated by a material initiator; (b) migration polymerization of a bifunctional oligomer in the presence of a trifunctional cross-linking agent.

formation of a polymer network as a result of migration polymerization of bifunctional epoxy or glycidyl oligomers in the presence of a trifunctional monomeric cross-linking agent. Note, however, that this is an idealized representation (initially considered by Carothers, Flory, and others), which must lead to a regular defectless network.

Analysis of the mechanisms of network formation suggested by different scientific schools [12, 13, 17, 20, 81] leads to an unambiguous conclusion that the overall process is microheterogeneous. Moreover, it is heterogeneous both in kinetic and structural aspects.

Modern concepts of oligomer curing were generalized within the framework of the scheme suggested by Alfred A. Berlin. This scheme is illustrated in Fig. 3.5. Since it was first described, this scheme has been the topic of permanent discussion. However, the discussion dealt with particular features and details of the scheme but did not interfere with its basics.

---

other similar molecules may in principle give rise to four network junctions.

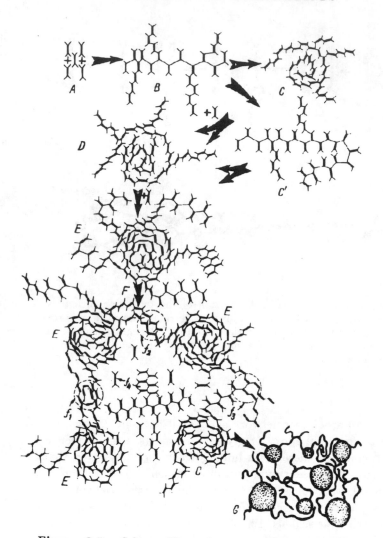

Figure 3.5. Scheme illustrating the formation and structural fragments of a network obtained by curing of unsaturated oligomers: (A) initial molecule of oligomer; (B) branched b-polymer; (C, C') coiled and anisometric shapes of the polymer; (D) cyclization; (E) gel nuclei (grains); (F) gelation: (f_1) entrapment of tie chains (loops), (f_2) entanglements, (f_3) chemical interaction, (f_4) unreacted molecules of the oliogomer and β-polymer; (G) completed microheterogeneous network.

The curing usually begins in "labile templates" (cybotaxises). In addition, spontaneous polymerization in other parts of the system may also take place. Because the local concentration of functional groups in the cybotaxises is higher than the concentration averaged over the entire volume, and orientation of the molecules in a "kinetically favorable order" is possible (see Fig. 2.10), the reaction rate is much higher in cybotaxises than in the other parts of the system.* Thus, we may speak of the kinetic microheterogeneity of the process. This means that, at each time instant, the polymerization proceeds at different rates within different microscopic volumes of the system. In its turn, this implies that the degree of conversion Γ is different in different parts of the reaction volume. Therefore, kinetic heterogeneity gives rise to structural heterogeneity. Structural heterogeneities perform as nuclei for the future heterogeneity of the entire system.

Let us recur to considering Fig. 3.5. Chain propagation increases the chainlength. During the initial stages, the oligomer predominantly enters into the reaction with only one of its bonds (provided that there is no kinetically favorable order). As a result, branched molecules with reactive pendant bonds are formed. Such reactive branched structures were called β-polymer. Prior to gelation, β-polymer is soluble in the reaction mixture composed of unreacted (free) oligomer. Depending on the chain flexibility and the magnitude of thermodynamic parameter describing the interaction with the medium, which are both the functions of the number of the addition events, β-polymer may acquire different conformations and even form a coil. At this stage, chemical reactions involving pendant bonds take place inside the coils (cyclization) or at their surface (front polymerization). The concentration of double bonds in the coil is higher than outside the coil, and, therefore, the rate of cyclization is initially higher than the rate of front polymerization. However, the pattern changes in the course of the process, and the front polymerization begins to dominate. This change is explained by the following causes. First, the number of double bonds capable of entering into the reaction gradually decreases, because the free oligomer containing the double bonds does not diffuse into the coil. Second, with an increase in thermodynamic incompatibility of the network and the free oligomer, the free oligomer is expelled to

---

* It is important to note that the concept of "kinetically favorable orders" made it possible to explain the abnormally high rates of the initial stages of polymerization, which were reported for some oligoester acrylates under certain conditions (for details, see [1, 13]).

the peripheral regions and outside the coil (microsyneresis [20]). Finally, increased local rigidity and steric limitations, first, reduce the accessibility of the pendant bonds and, then, make them inaccessible.*

Therefore, the local degree of conversion inside the coil Γ reaches its maximum level Γ_{max} long before the polymerization is completed in the other parts of the reaction volume and in the entire system.**

Hence, because of the kinetic microheterogeneity of the process of network formation, local phase separation may take place at the very early stages of the process, when the degree of conversion averaged over the entire volume is very low.

Further development of the process is associated with the growth of nuclei, although, according to [21], nucleation of the new propagation centers is still possible. However, this may only alter the size distribution of nuclei (see Section 3.7). Within the framework of the considered scheme, an increase in the size of nuclei results in that the chains that grow at its surface begin to entangle, entwine, and even chemically react via the pendant bonds. Concomitantly, the viscosity increases dramatically. The system loses fluidity, and this feature is identified with the gelation. Experimentally, this is observed as the first gel point. Depending on the nature of oligomer, mechanism of the curing reaction and the related conditions, this gel point may be observed at different overall degrees of conversion ranging from $\Gamma \leq 1$–2% (for oligoester acrylates) to $\Gamma \approx 20$–40% (for oligoepoxides) and even $\Gamma \approx 50$–60% (for urethane formation).

The second stage of the curing process is completed by the formation of at least two types of structural units, viz., nuclei in which $\Gamma = \Gamma_{max}$ and loose network interlayers between the nuclei [22], which still contain sufficient amounts of dangling bonds and of the free oligomer for the curing to proceed further.

During the third stage of curing, the polymerization rate is limited by diffusion. Polymerization proceeds as a front process. At $\Gamma \approx 75$–85%, the second gel point is observed. The grown nuclei agglomerate and the system becomes monolithic. This process is illustrated by the scheme in Fig. 3.6, which was suggested in [13]. According to [13], the evolution of the system results in phase inversion. Indeed, at the second stage, it is the nuclei that are dispersed in a loose network. In the cured system, it is the agglomerated nuclei (densely cross-linked

---

* This is one of the reasons why the degree of conversion never reaches 100%.

** Partly or completely polymerized coils are referred to as gel nuclei [17] or grains [13].

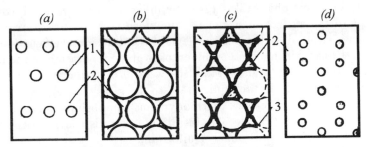

Figure 3.6. Evolution of phase organization during the cure of reactive oligomers: (a)–(d) different stages of the curing process; (1) nuclei, (2) loose network, (3) agglomeration of nuclei.

structures) that form the continuous phase, in which fragments of a loose network are dispersed. These fragments of a loose network constitute the defects of a cross-linked material.

In fact, the curing process considered above corresponds to the nucleation mechanism of phase separation. However, some authors [23 26] assume spinodal mechanism for phase separation during the formation of networks.

3.3. NUCLEATION AND SPINODAL MECHANISMS OF PHASE SEPARATION

Two mechanisms of phase separation in blended systems—nucleation and spinodal—are known. For polymer and oligomer systems, theoretical and experimental description of these mechanisms was reported in [24].

Stability of a system near the binodal is determined by the positive magnitude of the second derivative of Gibbs' free energy with respect to concentration ($\partial^2 G/\partial\varphi^2 > 0$). This implies that the system is unstable only when concentration fluctuations are large. Nucleation mechanism of phase separation is characteristic for this state of the system.

In the case of nucleation phase separation, a binary system, which is, in fact, a supersaturated solution of one component in the other component (in the distinct case considered here, a solution of β-polymer or a loose network in the free oligomer), begins to decompose by forming nuclei of the "daughter" phase in the "mother" phase. The growth of a new phase takes place by diffusion of the reacting compound to the surface of nucleus. The steadiness of nucleation and growth is ensured by the positive sign of the diffusion coefficient.

In the case when a system undergoing phase separation of the nucleation type reaches a glassy state (network formation may be treated as the transition from the liquid state to the glassy state), the resultant structure never reaches thermodynamic equilibrium, because the system gets frozen (fixed). The resultant system is characterized by incomplete (unachieved) phase equilibrium.

The system near the spinodal ($\partial^2 G / \partial \varphi^2 < 0$) is unstable even when the variation of the parameters of state is small (see Section 2.2). Therefore, the phase separation of supersaturated solution according to spinodal mechanism begins within the entire volume of the system without nucleation. According to Cahn [27], spinodal decomposition is a kinetic process involving spontaneous formation and permanent growth of the daughter phase in an unstable mother matrix. In the case of spinodal decomposition, diffusion flows are directed against the concentration gradient and the diffusion coefficient is negative, whereas in the nucleation mechanism, the diffusion flows tend to reduce the arising concentration deficiency. As a result, the so-called periodic or modulated structures are formed during spinodal decomposition.

Generally, the mechanism of liquid-phase separation is determined by the ratio between the rate of transition of a system to non-equilibrium state (W_{neq}) and the rate of its structural relaxation (W_{str}). At $W_{neq} < W_{str}$, nucleation mechanism is in effect, whereas, at $W_{neq} > W_{str}$, the system undergoes spinodal phase decomposition [7].

Rao and Rao [26] suggested the following scheme that identifies the principal differences between the two cases of phase decomposition:

Nucleation mechanism	Spinodal mechanism
Composition of the daughter phase is constant and does not change with time (at equilibrium)	The composition varies until equilibrium is attained
The phase boundary is sharp	The boundary is smeared
Size distribution of the particles of dispersed phase and distribution of the particles over the system are random	Size distribution of the particles and their distribution over the system are regular
Phase particles are spherical and are almost not connected to each other	Phase particles are not spherical and are frequently interconnected

3.4. RELATIONSHIP BETWEEN THE CURING KINETICS AND THE MULTISTAGE CHARACTER OF THERMODYNAMIC EQUILIBRIUM IN UNCURED OLIGOMER BLENDS*

The fact that the properties of materials based on oligomer blends depend on the prehistory of samples has been documented more than once [14, 28, 29]. For example, to obtain reproducible kinetic curves, a blend must be aged for a certain time prior to curing. For different blends this time varies from a few hours to several tens of days. This is related to a multistage character of the thermodynamic equilibrium in oligomer blends and to the fact that the time required to attain thermodynamic equilibrium in an oligomer blend is usually rather long; furthermore, the curing kinetics is sensitive to structural rearrangements that take place when the system moves towards equilibrium.

In [30–32], variation of the structure of an oligomer blend, was assessed in the initial polymerization rate terms of W_0 (this kinetic parameter is very sensitive to the structural organization of a liquid). As can be seen from Fig. 3.7, W_0 is a function of the exposure time τ_e elapsed from the moment of casting a liquid blend to the beginning of its curing. Moreover, in some cases, during the time period until the curve $W_0 = f(\tau_e)$ reaches a saturation level, W_0 grows, whereas in other cases it decreases. The time at which W_0 becomes independent of the exposure time (the curve reaches a saturation level) is denoted τ_e^{cr}. For two-phase systems, τ_e^{cr} is 2–2.5 times larger than for single-phase systems (Fig. 3.8).

These effects were explained within the framework of a cybotaxic model for the reorganization of supermolecular structure of an oligomer blend. In essence this explanation is as follows. As emphasized above, a kinetically favorable order in the arrangement of molecules in cybotaxises ensures that the initial rate of curing is higher in the cybotaxis than in the other parts of the reaction volume.** The greater the content of cybotaxises in a system and the larger their size, the higher is the overall W_0 observed in experiment for the entire system. Therefore, reducing the number and dimensions of cybotaxises

---

* We advise to recur to Sections 2.6–2.8 before reading this section.

** This is true only for those cybotaxises, whose lifetimes τ_l exceed the times of elementary events in a chemical reaction τ_{ch}. According to [33], for pure oligoester acrylates, these times are about 10^{-5}–10^{-6} s. In [92], the ratio between τ_l and τ_{ch} was assessed and compared with the resolving ability of different experimental techniques used for studying the curing kinetics.

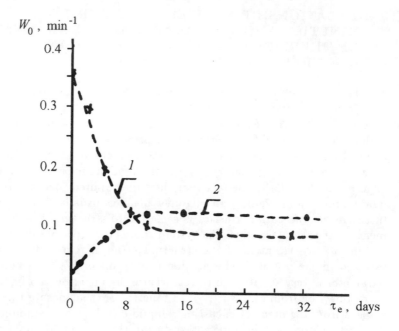

Figure 3.7. Plot of the initial rate of polymerization W_0 versus τ_e for blends of (*1*) butadiene–nitrile rubber and (*2*) *cis*-polyisoprene rubber with tetramethylene dimethacrylate (36.5 wt % and 40 wt %, respectively). Curing at 107°C, AIBN as initiator.

or reducing the lifetimes of fluctuations inevitably results in lower overall W_0; naturally, violation of the kinetically favorable orders in cybotaxises also reduces W_0. Each of the components of an oligomer blend is characterized by an inherent limiting (equilibrium) magnitude of the listed parameters of supermolecular structure. After mixing of the components, the equilibrium characteristic of each component is infringed, and the system tends to acquire a new equilibrium state. The averaging of the structural parameters of cybotaxises in an oligomer blend (i.e., attaining the equilibrium that conforms to the new thermodynamic conditions) takes certain time, which is limited by the magnitude of τ_e^{cr}.

The structural parameters of cybotaxises depend on the medium in which the cybotaxises appear. For oligomer blends with high thermodynamic affinity between the components, these parameters decrease at $\tau_e \leq \tau_e^{cr}$, because the components of such a system tend

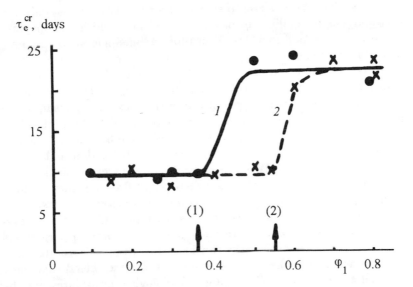

Figure 3.8. Plot of τ_e^{cr} versus φ_1 for blends of tetramethylene dimethacrylate with (*1*) *cis*-polyisoprene rubber and (*2*) butadiene nitrile rubber. The arrows show concentrations at which phase separation in the initial blends takes place.

to form a molecular dispersion and, therefore, W_0 as a function of τ_e declines (in Fig. 3.7, curve *1* was obtained for a system characterized by high affinity, $\chi_{12} = 0.012$). For poorly compatible systems the pattern is reverse, because the components tend to segregate (curve *2* in Fig. 3.7 corresponds to $\chi_{12} = 0.136$).

In two-phase systems, the attainment of equilibrium on a supermolecular level is complicated by interfacial mass exchange. In such systems, to averaging of the structural parameters of cybotaxises, taking place in each of the coexisting phases is accompanied by the diffusion-controlled coalescence of the particles of dispersed phase. Thus, the attainment of fluctuational association–cybotaxic equilibrium takes place under the conditions of permanently changing component concentration in the phases; naturally, this retards the attainment of "cybotaxic" equilibrium. This equilibrium becomes possible only after completion of the gradient-induced mass exchange between the phases. Therefore, τ_e^{cr} is significantly higher in heterogeneous oligomer blends than in the homogeneous ones (Fig. 3.8).

These features give rise to still another regularity in establishing thermodynamic equilibrium in oligomer blends, which is important for technology. As demonstrated in [30], τ_e^{cr} in single-phase oligomer blends is independent of the procedure used to prepare the samples, whereas this is not the case with heterogeneous blends in which the dispersity of a blend controls the magnitude of τ_e^{cr}. Obviously, the attainment of concentration equilibrium in the coexisting phases, which precedes the attainment of fluctuation–association equilibrium, is limited by diffusion. In heterogeneous systems, the diffusion coefficient is controlled not only by the temperature and composition, but by the interfacial area as well. When the specific surface area of the dispersed phase is higher, that is, when the dispersity of an oligomer blend is greater, mass exchange comes to a completion faster. The faster the concentration equilibrium in the phases is attained, the sooner the system comes to equilibrium with respect to the structural parameters of cybotaxises. Therefore, it is the procedures that ensure high specific surface area of the dispersed phase of oligomer blend that must be used to reduce τ_e^{cr}. The related technological procedures may be versatile. Recall also the possibility of increasing the dispersity of heterogeneous oligomer blends by using surfactants referred to in Section 2.7; that approach does not require additional power consumption.

3.5. RELATIONSHIP BETWEEN THE CURING KINETICS AND THE PHASE ORGANIZATION OF UNCURED OLIGOMER BLENDS

In Fig. 3.9, a phase diagram for the blend of tetramethylene dimethacrylate with cis-polyisoprene is compared with the plot of W_0 versus φ_1. Table 3.1 lists the values of W_0 that correspond to the single-phase state of two oligomer blends: to the "left" and to the "right" of the binodal. All W_0 values were obtained in uncured oligomer blends that reached thermodynamic equilibrium.

These data suggest that, in a single-phase state, W_0 is almost independent of φ_1. However, in the phase rich in oligomer (solution II in Fig. 2.1), absolute values of W_0 are 5–10 times higher than in the phase of oligomer solution in the polymer (solution I). In the region of transition from the single-phase state of the initial system to the two-phase state, the plot of W_0 as a function of φ_1 undergoes a sharp rise.

The causes that stipulate the differences in polymerization rates of oligoester acrylates in different phases of oligomer blends are related

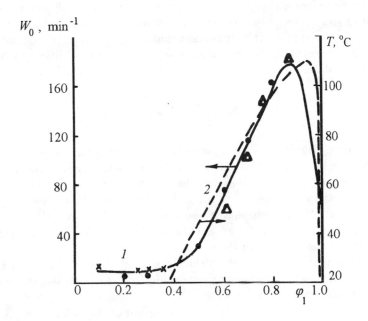

Figure 3.9. (*1*) Plot of W_0 versus φ_1 and (*2*) phase diagram for the blend of tetramethylene dimethacrylate with *cis*-polyisoprene; (×) and (•) correspond to experimental points obtained with initial blends prepared using different procedures, (△) calculated according to (3.1).

Table 3.1. Values of W_0 at different concentrations of reactive component φ_1 in single-phase blends of cis-polyisoprene with tetramethylene dimethacrylate (OB-1) and trioxyethylene dimethacrylate (OB-2)

System	Curing tempera-ture, °C	W_0, min^{-1}, at φ_1 (vol. fractions)							
		0.05	0.10	0.15	0.20	0.30	0.99	0.995	1.0
OB-1	100	0.09	0.12	–	0.14	0.16	0.97	1.05	0.32
OB-1	107	–	0.16	–	0.18	0.16	1.74	–	0.64
OB-2	107	–	0.10	0.13	–	–	–	1.58	0.88

to their different supermolecular organization; these will be elaborated on in the next section. Within the framework of the problem considered in this section, it is important that we may correctly calculate W_0 for heterogeneous oligomer blends in terms of the additivity scheme. The analysis reported in [31] led to the following relationship:

$$W_0 = W_{01} \frac{V_1}{V_0} \frac{\varphi_1'}{\varphi_1} + W_{02} \left(1 - \frac{V_1}{V_0} \right) \frac{\varphi_1''}{\varphi_1} \qquad (3.1)$$

Here, W_0 is the initial rate of oligomer polymerization in a heterogeneous system containing a volume fraction φ_1 of the reactive component; W_{01} and W_{02} are the initial rates of polymerization in the coexisting phases of solution I and solution II, respectively; φ_1' and φ_1'' are the corresponding concentrations of the reactive oligomer in these phases; V_0 is the total volume of the system; V_1 is the volume of the phase of solution I.

Comparison of the W_0 values obtained in experiment with the values calculated according to equation (3.1) evidences that they are in satisfactory agreement (Fig. 3.9).

Thus, knowledge of the phase diagram of oligomer blend and of the initial curing rate at two φ_1 values corresponding to the single-phase state of the blend allows a process engineer to calculate W_0 at any content of oligomer in the system, instead of carrying out a laborious routine experiment. Because in two-phase oligomer blends at $T = $ const the values of φ_1' and φ_1'' are independent of the total content of oligomer in the system (see Section 2.2), the reverse problem may be solved rather easily as well.

3.6. RELATIONSHIP BETWEEN THE CURING KINETICS AND THE SUPERMOLECULAR ORGANIZATION OF UNCURED OLIGOMER BLENDS

In the preceding section, a very important question, namely, why W_0 is greater in the phase of solution II than in the phase of solution I and higher than in the case of homopolymerization of "pure" oligoester acrylates, was left without answer. A trivial explanation resting on the theory of gel effect [9] is not applicable, at least because the viscosity of solution I is higher than that of solution II, whereas W_0 is higher. The answer to this question was found in [34–36], where the curing kinetics was examined for single-phase oligomer blends obtained by mixing a reactive oligomer (tetramethylene dimethacrylate) with nonreactive analogues, whose viscosity differed by two orders of magnitude.

Figure 3.10 shows the plots of W_0 versus the concentration of solutions. The following experimental observations are important for the analysis of the problem discussed:

(1) W_0 is a function of φ_1;

(2) for highly viscous blends at $1 > \varphi_1 \geq 0.7$, a pronounced maximum of W_0 is observed which gradually levels off as the viscosity of the second component decreases;

(3) for blends formulated of the components with close viscosities, W_0 monotonically decreases at $1 > \varphi_1 \geq 0.5$.

Let us examine these statements more closely.

First, for a given series of experiments, W_0 is normalized to the concentration of the reactive component, that is, its value refers to a single functional group (a methacrylic bond). According to the laws of classical chemical kinetics, W_0 expressed in this way cannot and must not depend on φ_1, provided that the second component does not alter the reactivity of the functional group (i.e., the second component does not show catalytic effect). Because the addition of a chemically inert oligomer to a methacrylate monomer does not affect the initial rate of its chemical transformations (Fig. 3.10, curve 4), the catalytic effect of the second component on the true reactivity of methacrylic bond on a molecular level cannot be admitted.

However, the fact that, for a mixture of oligomeric tetramethylene dimethacrylate with an inert analogue, W_0 is a function of φ_1 (Fig. 3.10, curves 1–3) may be related to reorganization of the supermolecular structure, which is inherent to blends of reactive oligomers, rather than to molecular rearrangements.

Second, for highly viscous oligomer blends, the extremum on the $W_0 = f(\varphi_1)$ curves must be the result of simultaneous occurrence of two competitive processes, one of which favors the growth of W_0, whereas the effect of the other is to reduce it.

The growth of W_0 in oligomer blends at $\varphi_1 > 0.7$, as compared to W_0 observed during the cure of "pure" oligoester acrylates, may be attributed to an increase in the number, size, and lifetimes of cybotaxises, which occurs when a more viscous component is added to an oligoester acrylate. However, this true premise, which agrees well with the idea suggested in [1, 13], does not reflect comprehensively the complex picture of the structure of cybotaxises formed in the blends.

Third, a monotonic decrease in W_0 on dilution of the system with an nonreactive component, as observed for oligomer blends with close viscosities of the components (Fig. 3.10, curve 3), was justified in [34,

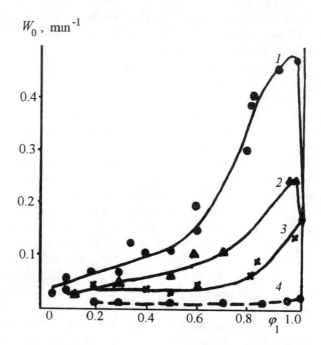

Figure 3.10. Plot of W_0 versus φ_1 for blends of tetramethylene dimethacrylate (MB) with (*1*) bis(dioxyethylene phthalate)-α,ω-diisobutyrate (IDP), (*2*) trioxyethylene diisobutyrate (TGI), (*3*) tetramethylene diisobutyrate (IB), and (*4*) a mixture of methyl methacrylate with IB.
Viscosity of components (cP): MB $\approx$ 5, IB $\approx$ 5, TGI $\approx$ 10, IDP $\approx$ 1000.

35]. We believe that this decrease is the result of "incorporation" of the nonreactive component into the structure of kinetically favorable order in cybotaxises (Fig. 3.11).

If this scheme is true, it is absolutely obvious that, even under the conditions when the size and the lifetimes of cybotaxises in such binary systems are the same as in pure oligoester acrylates, the number of elementary events in a chemical reaction in the initial time instant (and, correspondingly, W_0) is smaller in cybotaxises of type **B** than in those of type **A** (Fig. 3.11).

Therefore, one may easily explain the existence of extrema on curves *1* and *2* in Fig. 3.10. These extrema are due to a competition between the negative effect that the second component has on the

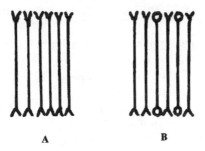

Figure 3.11. Schematic presentation of a cybotaxis of (A) an unsaturated oligomer and (B) its blend with a saturated analogue.

structure of kinetically favorable orders in cybotaxises and its positive effect as of a thickener (it increases the number of cybotaxises with lifetimes comparable with the characteristic times of the elementary events of the chemical reaction). Naturally, the higher the content of the second component, the higher is the number of cybotaxises of type **B**. Furthermore, the lower the viscosity of this component, the lower the content of cybotaxises of type **A**.

Thus, because in the phase of solution II (see Fig. 2.1) the concentration of the highly viscous polymer component is rather low ($\varphi_1 > 0.7$), the high initial curing rate in this phase is the result of its higher supermolecular organization.

Modern scientific literature lacks sufficient data to allow one to trace the quantitative relationship between the composition of the blend, its viscosity, temperature, etc. and the extent of supermolecular organization of the system and also reveal the quantitative correlations between the parameters of the cybotaxic structure of an oligomer blend and the kinetic constants of the curing reaction. These are the studies for the nearest future. However, even at the modern state-of-the-art in this field, the understanding of these qualitative correlations makes it possible for a process engineer to control the supermolecular structural order in an oligomer blend (and thus control the curing kinetics) by varying the technological parameters of the process (concentration, pressure, viscosity, temperature, etc.).

3.7. CURING KINETICS AND PHASE TRANSFORMATIONS IN OLIGOMER BLENDS

Regardless of the thermodynamic compatibility of the initial components, chemical transformations in oligomer blends always lead to phase decomposition of the system. Indeed, as a result of polyreactions, the molecular mass of at least one of the components of the blend increases. The higher the molecular masses of the blended components, the poorer their compatibility (with a few exceptions mentioned in Section 2.3). The higher the extent of polyreactions, the higher the molecular masses. Therefore, compatibility of the reaction products decreases with conversion Γ.

Figure 3.12 shows the phase diagrams obtained at different conversions in the curing of a blend comprising an oligomeric diene rubber with terminal carboxy groups and a bisphenol-A-based oligomer ED-20 (cured with m-phenylenediamine) [37]. In the initial state, this system forms a single phase at any ratio between the components. As the reactions progress, thermodynamic compatibility of the reaction products decreases, and, at $\Gamma > 30\%$, the compatibility of the mixture becomes limited. Further growth of Γ contracts the domain of compatibility. At $\Gamma > 55\%$, the branches of binodal are very close to the vertical axes; this suggests that the reaction products are totally incompatible at any composition of the reaction mixture.

Boundary conditions for compatibility of the curing products of oligomer blends are largely controlled by the nature of reactive component and the mechanism of polyreactions. In the example referred to above, phase decomposition under the specified curing conditions begins at $\Gamma > 30\%$. For acrylic oligomers, phase decomposition occurs at much lower conversions ($\Gamma < 1$–5%), and for allylic oligomers, at much higher conversions ($\Gamma \sim 55$–80%). In the general case, the rate of phase decomposition is controlled both by nature of the components of oligomer blend and by the curing conditions.

Quantitative data on the kinetics of phase transformations may be extracted by interpreting the plots $\Gamma = f(\tau)$ in terms of the Avrami model, which was initially suggested to analyze crystal growth. This approach was first applied to the analysis of network formation in [17]. Its validity was later verified in a number of independent studies [38, 39].

The essence of the suggested analysis is as follows. The kinetic curves are plotted in the coordinates of the modified Avrami equation

$$\Gamma = 1 - \exp(-k\tau^n) \tag{3.2}$$

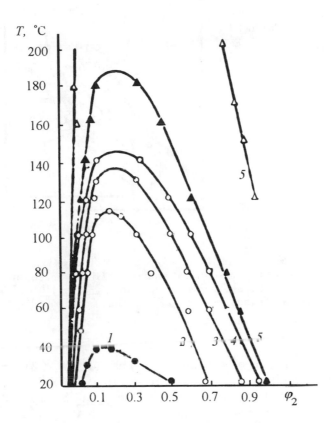

Figure 3.12. Phase diagrams for oligocarboxydiene–oligoepoxide blend at different conversions in the curing process: (*1*) 31, (*2*) 37, (*3*) 40, (*4*) 42, (*5*) 44, (*6*) 53% [37].

where Γ is the degree of conversion, $k \approx k_p/k_t$ is the specific rate of polymerization, defined as a ratio between the rates of chain propagation and chain termination, τ is the current time, and n is the index that depends on the polymerization conditions.

When the rate of a chemical reaction is much higher than the rate of phase separation, that is, when the reaction proceeds in a homogeneous medium, $n = 1$.

When the rate of a chemical reaction is lower than the rate of phase separation, that is, when the reaction is heterogeneous, $n \rightarrow 3$.

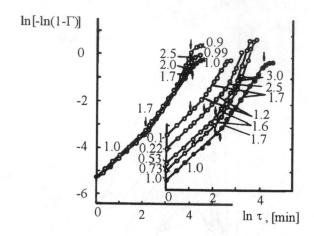

Figure 3.13. Curing kinetics for a blend of trioxyethylene dimethacrylate with its unsaturated analogue TGI in the coordinates of equation (3.2). The curing temperature T_{cure} was 110°C. The initiator was dicumyl peroxide (2%). See the text for explanations.

All variations of n within these limits ($1 < n < 3$) characterize the superposition of a homogeneous (in solution) and heterogeneous (near the interface) contributions to the overall polymerization rate. When $n = 0$, the reaction stops.

Thus, the pattern of the variation of n with t or Γ makes it possible to assess the kinetics of phase decomposition and the curing mechanism.

Figure 3.13 shows anamorphoses of the kinetic curves in the coordinates of equation (3.2) for the curing of trioxyethylene dimethacrylate blends with its saturated analogue [36]. In this figure, the values from 1.0 to 0.1 placed at the ends of the curves denote the weight fractions w_1 of oligoester acrylate, whereas the figures along the curves refer to the values of n at different stages of the curing. The arrows denote the bounds of these stages. Note that the viscosities of the blended components are equal and the blends are initially in the single-phase state.

The analysis of anamorphoses evidences that, in complete agreement with the scheme initially suggested by Berlin (Fig. 3.5), the polymerization starts and proceeds for some time (until certain values of Γ are attained) as a homogeneous process. At this stage, $n = 1$.

Then, n begins to decrease. This takes place at a certain critical magnitude $\Gamma = \Gamma_{cr}$, which, for the given oligomer blend, varies with w_1 from 1 to 4%. At this stage, n varies for different w_1 from 1.2 to 1.7. Therefore, at $\Gamma > \Gamma_{cr}$, a chemical reaction is partially localized at the interface that formed as a result of phase separation. As the process progresses, the concentration of reactive component decreases and, correspondingly, the overall reaction rate also decreases. At $\Gamma = \Gamma_{sl}$, the slow stage of the process begins; at this stage, n first increases approaching 3 and then decreases to 0. This means that the reaction first proceeds as a front reaction and then stops. According to the scheme shown in Fig. 3.6, phase inversion and agglomeration of nuclei take place at this stage. The value $\Gamma = \Gamma_{sl}$ corresponds to phase inversion. For given oligomer blends under the specified curing conditions, Γ_{sl}, varies from 50 to 80%.

Comparison of the anamorphoses for homopolymerization of trioxyethylene dimethacrylate (curve $w_1 = 1.0$) with those for the curing of this oligomer in the presence of its nonreactive analogue trioxyethylene diisobutyrate (the other curves in Fig. 3.13) leads one to conclude that there are no essential differences between the patterns that describe phase separation in these systems. Indeed, in all cases, curing is observed as a three-stage process: a homogeneous (at $\Gamma < \Gamma_{cr}$) process; the stage associated with the onset of phase decomposition at $\Gamma = \Gamma_{cr}$ (in this case, chemical reactions of two types—in the bulk of system and at the surface—occur); and, the final stage, at $\Gamma < \Gamma_{sl}$, when the reactions in the bulk are almost completely suppressed.

The effect of nonreactive component is confined to shifting the boundary values Γ_{cr} and Γ_{sl} and, correspondingly, changing the overall rates of chemical reactions at one stage or another. The extent of this influence is predominantly controlled by the initial content of the second component w_2. At $w_2 < 10\%$, the addition of nonreactive oligomer with the viscosity equal to that of oligoester acrylate has almost no effect on the kinetics of structural transformations during the initial stages of polymerization, but promotes phase separation at high conversions.* As a result, the conversion, at which phase inversion and monolithization take place, increases.

Increasing w_2 above 10% affects the kinetics of phase separation at any of the three stages. The quantities Γ_{cr} and Γ_{sl} increase almost proportionally to w_2. The value of n at the second stage of

---

* It is only phase transformations at the initial stages that we speak about. For the type of oligomer blends discussed, the rate of chemical interaction W_0 decreases at this stage (see Section 3.6).

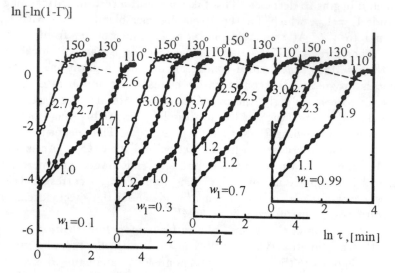

Figure 3.14. Curing kinetics for a blend of trioxyethylene dimeth-acrylate with PVC in coordinates of equation (3.2) at different T_{cure} and different contents of the oligomer w_1 in the starting blend. See the text for explanations.

the process decreases, that is, the contribution of heterogeneous reactions reduces and, correspondingly, the limiting degree of conversion in polymerization increases.

In addition to being controlled by the concentration of the second component, the kinetics of phase separation is also affected by the viscosity of this component, and, other conditions being the same, by the curing temperature. This is illustrated in Fig. 3.14, which shows the kinetic anamorphoses for the curing of PVC–trioxyethylene dimethacrylate blends of various compositions at different temperatures. The contents of the components were chosen so that, according to the phase diagram (Fig. 2.3), they covered the regions associated with the single-phase state of the system from the left ($w_1 = 0.1$ and 0.3) and the right of the binodal ($w_1 = 0.99$) and also the two-phase state ($w_1 = 0.7$) of the system.

Let us examine the effect of curing temperature. At 150°C, the initiated curing of a polymer–oligomer system begins as a heterogeneous process ($n = 2.5$–3.0) both in the single-phase and the two-phase states, that is, a rigid network is formed at the very beginning of the polymerization process. The situation is different at 110 and 130 °C. At these temperatures, the differences related to the phase state of a system are revealed.

In the phase of a solution rich in the polymer, where the rates of polymerization at the initial stages are even lower than those during the homopolymerization, the process is homogeneous (up to $\Gamma = 1$–6%) and phase separation does not take place. In contrast, in solution of phase II, where the oligomer is thickened by the dissolved polymer, and W_0 is 3–5 times higher (see Section 3.5), from the very beginning $n > 1$. Thus, in this solution phase separation begins earlier (at $\Gamma \leq 1$–2%) and proceeds faster.

In two-phase PVC–OEA systems ($0.33 < w_1 < 0.98$), phase separation begins simultaneously in both phases, but proceeds at a different rate in each of the coexisting phases (in solution I and solution II). Moreover, in each of the phases, the curing conforms to the regularities that are valid for single-phase systems. Therefore, the structural and kinetic parameters measured during the cure of heterogeneous systems are, in fact, averaged values. However, in contrast to the case with W_0, their averaging does not fit the additivity pattern.

Note another important feature that was deduced as a result of formalization of the curing kinetics in terms of Avrami model. As can be seen from Fig. 3.14, on one of the curves ($w_1 = 0.3$, $T_{cure} = 110°C$), the value of n at the second stage is greater than 3. It is not a misprint or a miscalculation and not even an experimental artifact. In [36], a set of n values which were greater than 3 and as high as 5.5 was obtained for different oligomer blends under certain curing conditions. High n values have been previously reported in the literature. For the case considered here, one may trace a feature similar to that reported by Mandelkern [40] for the crystallization, viz., sporadic formation of new nuclei. It is this feature that leads to n values that exceed the magnitude predicted by the Avrami model.

Thus, we have obtained experimental evidence that nucleation of new active sites may take place during the polymerization of oligomer blends, at some stage of the curing process. The theory of this phenomenon as applied to the curing of oligomer blends was elaborated by Rozenberg [21, 41, 42, 91].

From the "structural" viewpoint, the nucleation of new heterogeneity sites is associated with variation of the number and size of dispersed particles. Indeed, if the phase formation sites appear at random, the size distribution of the resultant particles is monomodal. This pattern is predicted by the theory of nucleation [43–45]. This pattern was also confirmed in a number of experimental studies [7, 21,

23, 24, 27, 46]. The width of the distribution depends on the nature of the components of an oligomer blend, their relative contents, and the curing conditions. If new particles are nucleated during the cure, the size of the new particles may significantly differ from that of the primary particles and, consequently, the final size distribution may be different from the monomodal Poisson's function. Apparently, it is this feature that gave rise to bi- and trimodal distributions reported in some publications [21, 47]. The sequential formation of nuclei was justified in [21, 41, 91].

The theory defines the condition for repeated nucleation as a supersaturation $\gamma = c/c_{eq}$ of the solution at a time instant t at which $d\gamma/dt > 0$. Here c and c_{eq} are the current and equilibrium concentrations of the dissolved compound. Thus, if the first derivative of oversaturation with respect to time is positive, the appearance of a new particle is possible. After the initial nucleation has occurred in a common polymerization reactions, the value of γ remains smaller than unity and, therefore, $d\gamma/dt < 0$. The situation is more complex in the case of network formation. Because $dc/dt = D_v d^2c/dx^2$, where x is the spatial coordinate of the dissolved compound near the growing particle, and $dc/dt = \beta \cdot \Gamma \cdot W$, where β is the coefficient depending on the degree (Γ) and rate (W) of polymerization, the rate of a decrease in c is proportional to D_v and there are no prohibitions on the change of the sign of the derivative $d\gamma/dt$. The probability of nucleation J_t at a time instant t is described by a complex functional relationship between J_t and variation of the free energy of the system upon the formation of a critical nucleus, structural parameters of the nucleus (radius, number of particles inside the nucleus and at its surface), density and molecular mass of the medium in which the nucleus is formed, interfacial tension, temperature, and, naturally, the magnitude of γ [91]. The last quantity is determined from the concentration profile of dissolved compound in the space between the particles of the first and subsequent nucleations.

Verification of this theory was based on the results of a complex experiment studying the curing of rubber–epoxide blends [21, 37]. Correlations between thermodynamic and kinetic parameters of curing and the morphology of final structure were revealed.

Figure 3.15 illustrates the variation of the curing rate W, coefficient of mutual diffusion of the components D_v, and their limiting solubility c_{eq} with the degree of conversion in polymerization Γ for rubber–epoxide oligomer blends based on diglycidyl resorcinol ester (DGR) cured by 2,6-diaminopyridine (DAP). The rubbers used were the copolymers of butadiene with 8 and 14% of acryloni-

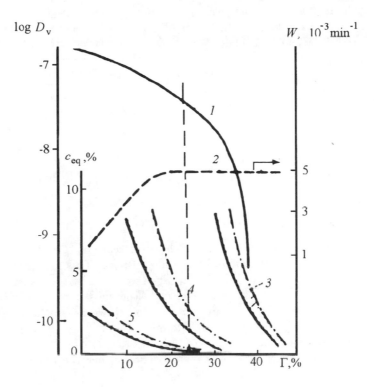

Figure 3.15. The plots of (*1*) logD$_v$, (*2*) W, and (*3–5*) c_{eq} versus
Γ for the blends of epoxide oligomer ED-22 with (*1–3*) PEF-3A,
(*4*) SKN-8-KKG, and (*5*) SKN-14-KKG.

trile end-capped by carboxylic acid groups (SKN-8- KKG and SKN-
14-KKG, respectively) and oligotetrahydrofuran with urethanepoxide
end groups (PEF-3A). The data deduced from these plots were used
to calculate the probability of repeated nucleation. These calculations revealed that, at a distance $0.5l$ between two growing particles
of the primary nucleation that occurred at an instant t, the probability of a secondary nucleation site to be formed dramatically (by several
orders of magnitude) decreases because of the migration of reactive
component to the surface of the growing particle and, consequently,
the reduction of concentration c at the site of presumed nucleation.
However, the same calculations revealed that at a certain time instant
t_2, J_{t2} increases, because of a decrease in D_v and a simultaneous de-

Table 3.2. Structural parameters of the cured blend DGR + PEF-3A obtained at different curing rates ($T_{cure} = 120°C$; the curing rate was controlled by adding a catalyst)

W, 10^{-4} s^{-1}	Volume fraction of heterophase, %	Distribution type
0.8	3.0	Monomodal
1.2	2.7	Monomodal
4.1	1.5	Monomodal
6.4	0.6	Bimodal
8.3	0.5	Trimodal
12.1	0.3	Bimodal
14.8	0.2	Monomodal

crease in c_{eq} that take place with an increase in Γ, and the derivative $d\gamma/dt$ changes its sign and becomes positive. This means that certain conditions for repeated nucleation are created. As a result, new particles appear. Their growth retards the subsequent nucleation near the particles of the second generation. However, to the time instant t_3, the same reasons as considered above restore the conditions for nucleation of the particles of the next generation. This oscillatory phase formation comes to a stop when either the molecular mobility is completely frozen ($D_v \rightarrow 0$, because of gelation or isothermal vitrification), or the free oligomer is completely consumed ($c \rightarrow 0$).

Under real curing conditions, the size distribution pattern is controlled by many factors, of which the curing rate is the most important. This statement is illustrated by Table 3.2, which demonstrates that, as the curing rate increases, the particle size distribution first changes from monomodal pattern to the bimodal and then the trimodal patterns to be finally "degenerated" to a monomodal distribution again.

The experimental data discussed in this section were interpreted as resulting from phase separation that takes place according to the nucleation mechanism. For the systems PVC–oligoester acrylate and liquid rubber–oligoepoxide, which were discussed above, the nucleation mechanism was confirmed in structural studies [21, 29, 36].

For some other oligomer blends, spinodal phase decomposition was assumed to occur during the cure [5, 7, 24, 46, 48]. For example, SAXS studies of IPNs and semi-IPNs based on segmented and cross-linked polyurethanes revealed periodicity in the variation of composition with a period of about 30 nm. This periodicity was associated with the spinodal mechanism of phase separation during the structure formation by the two network components. The lack of inflections on the kinetic anamorphoses is also referred to as a proof of the spinodal decomposition [24].

In a number of publications [7, 29, 49], the occurrence of both mechanisms of phase separation during the cure of oligomer blends was claimed. Both mechanisms can come into play at different stages of curing or simultaneously, but inside the different phases of an initially heterogeneous system.

However, one must keep in mind that, regardless of the mechanism of phase separation, the final structure formed as a result of curing of an oligomer blend is always characterized by incomplete phase separation and separation of the components, that is, the separation that must have taken place owing to thermodynamic reasons is never completed. The degree of incompletion is controlled by many factors but, primarily, by the curing kinetics, that is, by the rate of suppression of translational mobility of those structural elements of a system that, according to thermodynamics, must have separated and alienated from each other.

3.8. STRUCTURE FORMATION DURING THE CURE OF POLYMER–OLIGOMER BLENDS

In the discussion above we emphasized that, because of the significant difference in chain propagation rates inherent to each of the components during the cure of an oligomer blend, macroscopic curing may be treated as a reaction of one of the components in a polymer matrix formed by the other component, viz., the one that has formed faster and earlier. Therefore, curing of polymer–oligomer blends, in addition to being an important subject per se, is also interesting as a model for the curing of any other oligomer blend in which the propagation rate constants for the components are significantly different.

3.8.1. Curing of Single-Phase Blends

Two types of single-phase polymer–oligomer systems are distinguished (see Fig. 2.1): solutions rich in polymer (solution I) and solutions rich in oligomer (solution II).

Curing of polymer-rich solutions. In the first-order approximation, this process may be described by a scheme depicted in Fig. 3.16. As in the case of homopolymerization of the oligomers (Fig. 3.5), the process starts with fast-initiated polymerization in the cybotaxises, in which the concentration of functional groups is higher than the average concentration over the entire system. It is in the cybotaxises that the primary sites for the future heterogeneity of the cured composite are nucleated. The structure formation involves intramolecular cyclization, chain branching, and intramolecular interaction between the fragments of a loose network. As a result, microgels, which in fact represent a network swollen in the other components, are formed.

Then, other oligomer molecules that are dissolved in the polymer are involved into the process. They take part in front polymerization (by diffusing to the surface of primary nuclei) and in the generation of new branching sites in other parts of the reaction volume.

Naturally, in the case of front polymerization, the chain propagation rate constant is limited by the diffusion of reactive molecules to the active site even to a greater extent than during the homopolymerization (in oligomer blends, the diffusion is much slower because of the high viscosity of the medium). However, the same reason causes the chain termination rate constant to decrease also. Therefore, the influence of solution viscosity on the rate of network formation is controlled by the competitive contribution of the two factors with opposite effects on the elementary reaction stages.

Generation of new branching sites is primarily controlled by the rate of initiation. If the rate of decomposition of an initiator is comparable to or lower than the polymerization rate, the active sites may appear continuously as the initiator decomposes. If the chain propagation is not limited by the rate of initiation, oscillatory nucleation mechanism described in the preceding section may come into effect.

As the molecular mass of a network and the extent of intramolecular cyclization increase (i.e., as the degree of conversion Γ increases), mutual solubility c_{eq} of the components drastically decreases. As a result, thermodynamic equilibrium that has been previously attained is violated. The derivative of local supersaturation becomes positive, and phase separation starts on a microscopic level.

At this stage, differences between the curing of an oligomer blend and homopolymerization of an oligomer are discernible. In homopolymerization, the polymerized system is a mixture of variable composition, which contains network fragments with different degrees of cross-linking, β-polymer, and the free oligomer that has not yet entered the reaction, that is, all components of the mixture are of the same chemical origin. When a polymer–oligomer blend is cured, in addition to the products listed above, the reaction mixture also contains a linear polymer, which is usually chemically different from the

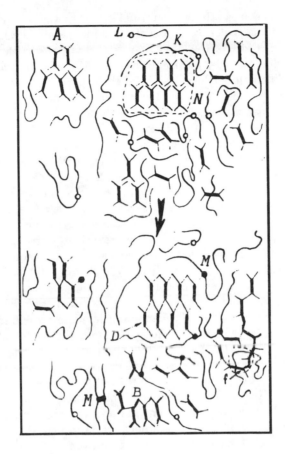

Figure 3.16. Scheme illustrating the curing of a single-phase polymer–oligomer blend: (A) molecules of oligomer; (K) molecules of polymer; (N) cybotaxis; (L) potentially active center on a polymer molecule; (B) β-polymer; (D) cross-linked polymer; (M) covalent bond; (f) entanglements, loops, entrapments, etc.

other components. In its turn, this feature controls the specific features of phase separation in oligomer blends on a microscopic level. First, the presence of linear macromolecules reduces c_{eq} and thus accelerates the onset of nucleation or phase separation by any other mechanism. Second, during microsyneresis, when the mobility in the microgels becomes strongly retarded (local isothermal vitrification), some macromolecules of the matrix polymer—and to a much greater extent than those of the β-polymer and, moreover, the molecules of the free oligomer (for macromolecules of the matrix polymer translational diffusion coefficients are much lower than those for the free oligomer)—fail to be expelled beyond the interface between the phases. These macromolecules remain embedded in the microparcels of the incipient cross-linked aggregate.

Such structures may be considered as a mixture of cross-linked and linear macromolecules of different structures characterized by incomplete separation of the components.* They evolve in the macroparcel of the system as isolated highly disperse particles (in Russian-language literature they were conventionally called Ångström-size particles [29].** From the viewpoint of thermodynamics these particles are non-equilibrium; however, under real conditions they are rather stable, because in these particles translational diffusion is completely suppressed.

From the curing scheme considered it follows that the most densely cross-linked (and the most compactly packed) part of the network is located in the center of a nanoparticle. The cross-linking density gradually decreases with the distance from the center of a particle. At the periphery, the particle represents a loose network containing entrapped macromolecules of a linear polymer. The space between the particles (the part of the system that is characterized by the loosest packing) is occupied by the macromolecules of a polymer that are mixed with loose network fragments and the remaining unpolymerized oligomer. In this way, a gradient structure—both with respect to the cross-linking density of a network and the molecular composition (relative contents of the network and linear polymer)—is formed.

The interaction between the unlike components is worth a special discussion. One cannot exclude chemical interaction between a linear polymer and an oligomer or the products of its reactions. However, this pathway is only a particular case. Formation of covalent bonds between the molecules of a polymer and oligomer, or the grafting of network fragments—products of oligomer curing—to a linear macromolecule are possible only under special conditions, which include the presence of appropriate functional groups or unsaturated

---

* In the general case, the volume fraction of the molecules of the free oligomer may be ignored, because it is incommensurably lower.

** This term is used to describe the scale of geometric dimensions of these particles only. In this respect, it is similar to the term "nanoparticle" [50] introduced later.

bonds in a macromolecule (see Table 1.1), the corresponding initiation or catalysis, special temperature regimes, etc. "Random" chain transfer or termination on a polymer is also possible. These potential pathways for the development of the process are reflected in the scheme presented in Fig. 3.16. However, the analysis of experimental data leads to unambiguous conclusion that the assumption that chemical interaction between unlike components during the cure of oligomer blends is always present, which was once popular in the 70s [51–56], is not generally true. Figure 3.17 illustrates the hypothetical structures that may form during the cure of polymer–oligomer blends which are currently discussed in the literature. To understand the nature of interaction in each distinct case, a scrupulous analysis of the possible mechanisms of relevant chemical reactions must be carried out. Presently, one may claim that chemical grafting of oligomer and the products of its reactions to a polymer does take place only in a number of rather well-studied processes, viz., radiation-induced curing of PVC–oligoester acrylate blends [61–65], peroxide vulcanization of the blends of nitrile rubbers with polyfunctional oligomers [6, 29, 68, 69], and some other [2, 14, 63–65].

Chemical aspects of the interaction between the components of oligomer blends during the cure are rather obscure and require special consideration, which is beyond the scope of this monograph.

Spontaneous fluctuations in cybotaxic liquids, random generation of branching sites, the dependence of their number on the concentrations of initiator and inhibitor, and on the temperature of the process, diffusion limitations at the stage of chain growth, etc.—these features determine the formation of disperse particles, whose size distribution is in most cases described by a monomodal function, although, as emphasized above, polymodal distribution is also possible.

Experimental studies demonstrate that the mean size of disperse particles that are formed during the cure of single-phase polymer–oligomer blends varies from several tens to several hundreds of angstroms [21, 24, 29, 36, 71]. However, formation of larger particles (as large as several thousands of angstroms) was also reported [67, 68]. The mean distance between the particles is about tens to hundreds of angstroms [7, 24, 70].

As demonstrated in electron microscopic studies, nanoparticles are generally globular, and this agrees with the nucleation mechanism of phase formation [21, 29, 36]. However, particles of distorted shape and with blurred boundaries between them (which is one of the features associated with spinodal decomposition) were also reported [7, 24, 48].

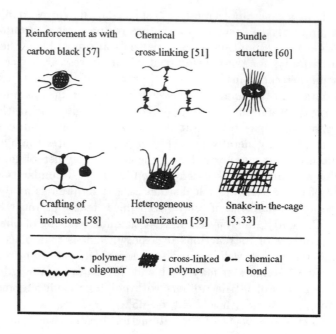

Figure 3.17. Schematic presentation of the structures that can form during the cure of polymer–oligomer blends.

When analyzing the macroscopic structure of cured oligomer blends, one must take into account the effect of the medium on the molecular composition of the structure being formed (incomplete separation of the components, gradient, etc.) and also on the parameters of other hierarchic levels. For example, physical interaction of the growing branched molecule with the matrix (their miscibility, polarities, thermodynamic affinity, etc.) determines the conformations that are allowed for the growing chain (e.g., whether the extended-chain or the coiled conformation is favorable). In its turn, the conformation controls the steric accessibility of the functional groups in intramolecular cyclization, branching, and front polymerization, that is, it controls the topology and morphology of the network formed. In the final account, this may result in additional ordering (or, conversely, disordering) of the structure of individual particles or of the entire system. Hence, in spite of the fact that in most of the reported cases the structure of cured oligomer blends was globular, supermolecular structures of different shapes (e.g., band structures and even spherulite-shaped structures) were also observed in some electron microscopic studies [26, 69].

Curing of oligomer-rich solutions. Curing in solution II is generally similar to how it proceeds in solution I. The differences that are observed are associated with the fact that the concentration of polymer in solution is low. The viscosity of such solutions is by several orders of magnitude lower than in solution enriched with the polymer, whereas the curing rate is 5 to 10 times higher. The structural consequences of these features are obvious. The onset of structure formation is shifted to lower degrees of conversion. Under identical initiation conditions, the mean size of nanoparticles that are formed as a result of the curing of phase II is 1.5–2 times larger than of those formed in phase I. The particle size distribution is likely to broaden, and the distribution may even become bimodal or trimodal (see Section 3.7). The body of a particle (according to the terminology used in [13]) is depleted of linear polymer. The content of the polymer in the space between the globules is also much lower; the composition of interglobular medium is closer to that of the "interparticle matter" in the cross-linked polymer obtained by homopolymerization. Nevertheless, network products of the curing of solution II always contain linear molecules of the second component; however, their content is lower than in the cured structure of solution I.

Thus, in spite of the fact that initial oligomer blends are, in both cases, single-phase and the constituent components are chemically similar (this feature explains the fact that the dimensions of the particles formed as a result of curing are of the same order of magnitude), some parameters of their structure differ significantly. These differences primarily involve the molecular composition of the cured products and the density gradient of the networks. The differences reported in [29] are so dramatic that they may be associated with the structures of different types.

The suggested structures of nanoparticles formed as a result of curing of solutions I and II are illustrated in Fig. 3.18. The radial variation of the cross-linking density is also shown.

3.8.2. Curing of Two-Phase Blends

In two-phase systems, which are in fact emulsions of the solutions I and II, curing occurs simultaneously in the coexisting phases and does not differ from the processes considered above for each of the single-phase systems. As a result of curing, heterogeneous composites with

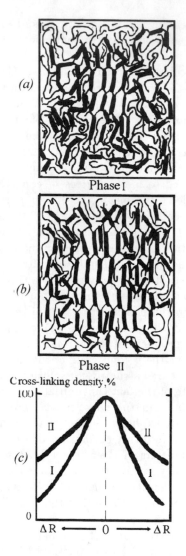

Figure 3.18. Schematic presentation of the structures of nanoparticles: (*a*) Curing of the solution of oligomer in polymer (I) and (*b*) of the polymer in the oligomer (II); (*c*) Radial distribution for network density; ΔR is the distance from the center: element >—< denotes the molecule of oligomer within a network fragment; element ∿∿ denotes the molecule of polymer.

structural elements inherent to each of the cured phases are formed. Therefore, the macroscopic structure of these composites must be considered as a mixture of two microstructures that are formed during the cure of the solution of polymer in oligomer and the solution of oligomer in polymer.

However, it is an essential feature of the composites based on two-phase systems that the distribution of these two structures in the final material is controlled by the morphology of the initial blend. The size of the droplets of a polymer–oligomer emulsion and their distribution in the initial (prior to curing) blend are determined by the nature of constituent components, content of oligomer, and the prehistory of a system (see Section 2.7). At the start of curing, the process occurs in the dispersion medium and within each droplet. When the dispersed phase is a solution rich in oligomer, fast polymerization in this phase (W_0 is 5–10 times higher than in solution I) fixes the size of droplets (Fig. 3.19). In the final structure of a composite, rather large disperse inclusions, which were conventionally called "micron-size" particles are observed [29].* Naturally, micron-size particles are heterogeneous. They are composed of compactly packed agglomerates of nanoparticles. By optical microscopy, micron-size particles are imaged as solid pieces with sharp boundaries [6, 68]; electron microscopy and SAXS allow the constituent nanoparticles to be also observed [29, 36, 61, 73]. According to [60], the structures of this kind may be classified as a "phase-in-phase" structures.

Thus, the difference in colloidal organization of composites obtained by curing single-phase and two-phase polymer–oligomer blends primarily pertains to the fact that the two-phase systems reveal an additional level of dispersity, that is, they contain micron- size particles. Indeed, when single-phase polymer–oligomer blends are cured, only nanoparticles are formed. In the case of a solution of oligomer in the polymer, nanoparticles of the first type are formed, whereas in the case of a solution of the polymer in oligomer, these are nanoparticles of the second type. When two-phase oligomer blends are cured, disperse inclusions of three types are formed: (i) nanoparticles of the first type, they are formed in the dispersion medium; (ii) micron-size particles, which result from the curing of the disperse phase; (iii) nanoparticles of the second type, they constitute the body of micron-size particles.

Naturally, if the composition of the initial blend is such that the continuous phase is the solution II, whereas the dispersed phase is that

---

* This term is used to refer to the geometric dimensions of the particles only (cf. footnote on p.116).

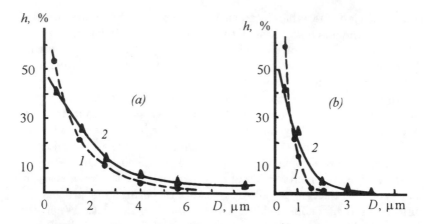

Figure 3.19. Particle size distributions for (*a*) (ethylene-propyle-ne rubber)–α,ω-dimethacryl-bis(glycerol)phthalate, (*b*) (SKD butadiene rubber)–α,ω-dimethacryl-bis(trioxyethylene)phthalate. (*1*) Prior to curing; (*2*) after curing. Data of optical microscopy.

of the solution I, micron-size particles are composed of nanoparticles of the first type, and nanoparticles of the second type are formed in the dispersion medium.

From the technological viewpoint, it is important that curing begins in a blend with the phase organization that has been established during the preceding stage of a process. In the case of single-phase blends, microheterogeneity results from the curing, whereas in the case of two-phase blends, the degree of dispersity is controlled by the morphology of the initial blends. The possibilities to regulate the structure of "crude" oligomer blends are discussed in Section 2.7.

The concepts discussed above are reflected in the scheme presented in Fig. 3.20 [29]. This scheme was elaborated for polymer–oligoester acrylate (OEA) blends and is also applicable to other oligomer blends. This scheme is rather illustrative and does not require wordy comments. Here we confine ourselves to emphasizing some technologically important features. In single-phase blends under identical conditions, the number of nanoparticles in the cured composite increases as the content of oligomer in solution increases. In two-phase oligomer blends, it is the number and dimensions of micron-size particles that increase in the cured composite with an increase in the oligomer content. Further increasing the content of oligomer leads to phase inversion in the initial blend, and the variation of morphology

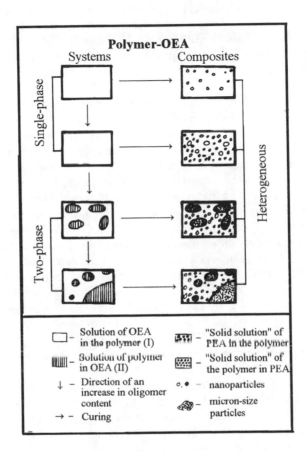

Figure 3.20. Schematic diagram illustrating evolution of the morphology of composites during the cure of single-phase and two-phase polymer–oligomer blends.

of the cured composite with the composition of oligomer blend follows a reverse pattern.

3.9. THE STRUCTURE OF CURED OLIGOMER BLENDS

The published experimental data illustrate versatility of the structures formed as a result of curing different classes of oligomer blends. The differences in the structure and properties of cured composites are primarily associated with the differences in the nature of constituent components. This is quite natural. Indeed, the properties of materials obtained by curing epoxide or phenol-formaldehyde oligomers are essentially different from the similar properties of epoxy–rubber composites [21].

Generally, factors that control the variety of structures in the cured oligomer blends may be distinguished into obvious factors, which were always taken into account by process engineers formulating compositions and selecting the curing regimes, and those that are not as obvious and are usually ignored in process design. However, as suggested by the experience of recent years, factors of the second group significantly affect the structure of composite materials.

The first group of factors involves the nature and functionality of the constituent components of a blend. It is these factors that determine the possible differences in the mechanism and kinetic features of curing. The ratio between the contents of the components is also a factor of this group. In the simplest case this ratio controls, for example, whether a rubber is modified by an epoxide oligomer or the epoxide network is modified by the rubber. In some cases, the procedure used to obtain the material is decisive. This statement is illustrated by the different structures of IPNs (similar in constituent components and composition) prepared by simultaneous and consecutive curing.

Factors that are not obvious involve thermodynamic affinity between the components, phase organization of the system that has been established prior to curing, degree of orientational order in supermolecular structures, and other variables that can be taken into account by adjusting the processing regimes. It is also important how far the initial blend is from the equilibrium state.

Note that the chemical nature of the components is of primary importance because it determines the thermodynamics and kinetics of the formation of the structure of composites. However, ignoring the kinetic and thermodynamic regularities may ruin the potential advantages related to the chemical structure of the components.

Whatever the causes that lead to different structures of cured oligomer blends with different prehistory, obtained from different components and in different regimes, there are always some general features in the structural organization of composites related to the fact that they are multicomponent systems.

The structure of cured oligomer blends has been the subject of controversy for many authors [2, 5, 7, 14, 21, 33, 51, 57–61, 68, 71]. However, after eliminating the ambiguities in terminology used by different authors to denote structural elements and the related processes and omitting some unimportant details, the views on the structure of materials obtained by curing of polyfunctional oligomer blends may be systematized to reveal some common features. Further on we discuss these features in some detail at different levels of structural hierarchy.

Molecular level. Naturally, this level is associated with the nature and functionality of cured composites. However, in any case, the final structure of the cured products may be described as a blend, either of two cross-linked polymers, or a cross-linked polymer and a linear polymer. Such blends are thermodynamically incompatible, and the constituent components of such a blend must have undergone separation. However, because of almost completely suppressed translational mobility, the structure of a blend does not change with time.* The degree of mixing of different components (incompletion of separation of the components or an inverse parameter, the extent of segregation of the components is controlled by the ratio between the rates of network formation and phase separation, that is, it depends on the stage of curing at which the system loses its ability to undergo microsyneresis.**

At very high extents of curing or when spinodal decomposition takes place, a molecular level of mixing is sometimes attained, that is, dispersion is fixed at a macromolecular level and, therefore, the extent of segregation of the components tends to zero ($\alpha \rightarrow 0$).*** However, in most cases, even for the initially single-phase systems, curing leads to structures characterized by nonuniform distribution of the components throughout the volume. This feature is a consequence of the local character of phase separation, which takes place with different rates in different regions of a macrosystem. Therefore, macroscopic structure of such cured systems is described by different degrees of mixing of the components throughout the volume and by the concentration gradient for this components in local microscopic regions, which is observed on the nanoparticle-scale. The composition

---

* The reasons for the suppression of mobility may be different. For example, molecular mobility may be suppressed as a result of isothermal vitrification, entanglements, formation of clathrate or snake-in-the-cage structures and even chemical grafting.

** The extent of segregation is calculated from SAXS data as $\alpha = \Delta\rho_1^2/\Delta\rho_c^2$, where $\Delta\rho_c^2 = \varphi(1-\varphi)/(\rho_1-\rho_2)^2$, ρ_1 and ρ_2 are the electron densities of different microscopic regions, φ is the volume fraction of one of the components, $\Delta\rho_1^2$ is the mean square fluctuation of electron density in the experiment [24, 72].

*** Obviously, such "solid solutions" are non-equilibrium.

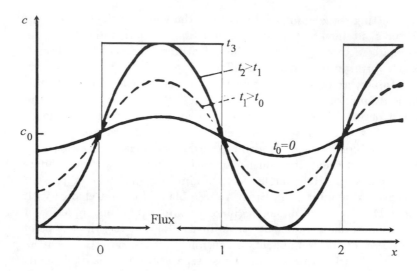

Figure 3.21. Spatial variation of the amplitude of composition fluctuation at different time instances t [26] (x is the spatial coordinate).

of nanoparticles differs from the average composition of a system and a varies from 1 to 75% [7].

The molecular structure of cured oligomer blends can be discussed in alternative terms. For example, by analogy with sinusoidal function for the amplitude of composition fluctuation during spinodal decomposition (Fig. 3.21), one may assume that during the cure of an oligomer blend that proceeds for a certain period of time t (from the onset of chemical reaction to the maximally attainable, under the selected conditions, conversion of functional groups), separation of the components stops at a certain time $t = t_1$ or t_2. It is this time that determines the molecular composition of a phase.

Some remarks are appropriate at this point of discussion. As emphasized by Lipatov [23], the concept of concentration fluctuation as it was used in [26] applies to single-phase systems only. However, these structures are not homogeneous. Indeed, as the curing progresses, they acquire some features of a heterogeneous system but do not become truly heterogeneous because the features that are assumed sufficient to classify a system as heterogeneous are lacking (see the definition of a phase in Section 2.1). Shilov and Lipatov [24] called these structures phasons. Phasons are a particular case of a dissipative structure provided that phase decomposition is considered within the framework of non-equilibrium thermodynamics (thermodynamics of irreversible processes [74, 75]). In contrast to equilibrium structures, which may be homogeneous and infinite in size, dissipative structures are heterogeneous and are described by characteristic dimensions. The specific feature of phason structures that makes them different from dissipative structures formed during the crystallization or vitrification is the formation of regions of two types, viz., those enriched with one and those enriched with the other component.

Let us emphasize one ambiguity in the analysis presented above. According to [7, 24], phason structures are formed only as a result of spinodal decomposition. However, as demonstrated above and as follows, for example, from Fig. 3.18., kinetically stable structures with incomplete separation of the components, regardless of how you call these structures, can be formed during the nucleation growth of a phase also.

When the initially two-phase oligomer blends are cured, the differences in concentrations of the solutions of the coexisting phases (solutions rich in and depleted of one of the components) and the distribution of the droplets of the dispersed phase (one of the solutions) in the dispersion medium of the other solution specify the discrete distribution of the cured microscopic regions with a higher content of the reaction products of one component or the other in the entire system. In other words, in those parts of the system that, prior to curing, represented the dispersed phase and the dispersion medium, nanoparticles of different types are formed. These particles are different from each other in their composition; their composition is also different from the mean composition of the system.

Topological level. For cured oligomer blends, the topology of mixed networks (i.e., concentration of network junctions, their functionality, relative amounts of the junctions of chemical and physical origin, length of chains between network junctions and MMD of these chains, number and complexity of ring structures, content of dangling chain ends, content of pendant bonds or functional groups, etc.; these parameters are referred to each of the networks) or the topology of the blends of cross-linked and linear products is characterized by the nonuniform distribution of like fragments over the volume of a system.

Like the molecular level, the topology of a system is controlled by the nature and functionality of the initial components of oligomer blends and the curing kinetics. Thermodynamic affinity between the components (which determines the configuration of growing chains and, consequently, the degree of intramolecular cyclization) and the procedure used to prepare a material (e.g., simultaneous or consecutive IPNs) are also essential. Important roles also belong to MMD and FTD of the initial components (these parameters specify the distance between network junctions and the branching of chemical junctions), etc.

In a comprehensive monograph [12], Irzhak *et al.* summarized the theoretical and experimental data on the regularities that describe the formation of topological structure in networks obtained by polymerization or polycondensation of individual oligomers.

Further development of these studies [84, 85, 88–90] revealed that the topological structures of amorphous polymers formed by free-radical polymerization of oligomers and by condensation polymerization are very different. These differences are primarily associated with the fact that the supermolecular structure of polymers is controlled by the ratio between the rates of initiation and chain propagation which specifies the localization of the process of polymer formation (structural heterogeneity). In addition, these rates are different for polymerization and polycondensation.

Structural and kinetic analyses reported in [85, 89, 90] made it possible to give a quantitative description of the formation of topological structure of linear and cross-linked polymers prepared by curing of various oligomers. The analysis rested on the concept of "bond blocks" [88] and took into account the equations that describe the growth of molecular chain for arbitrary kinetic constants and the equations describing ring formation. The final description involved MMD, bond blocks, interjunction spacings, number of cycles, etc. as parameters. In some cases, calculated data were in satisfactory agreement with the experiment.

Approaches to network topology developed for the case when a network is formed by an individual oligomer [12, 88–90] are also applicable to topological analysis of blended systems. However, in the case of blended systems, these approaches become much more complicated. Unfortunately, the published quantitative data on topological parameters of cured oligomer blends are obviously insufficient to allow serious generalizations to be made.

Present-day knowledge gives reliable evidence of the existence of the cross-linking density gradient in local (on the nanoparticle scale) microscopic parcels of a system and periodicity in the cross-linking density of the elements of network on the scale of micron- size particles in the structures obtained by curing initially single-phase or two-phase polymer–oligomer blends [29, 36]. These features are illustrated in the schemes shown in Figs. 3.18 and 3.20.

One may also claim the existence of the cross-linking density gradient inward a material prepared by non-equilibrium swelling of network polymers in the "native" or "alien" oligomer and subsequent curing of this oligomer (so-called gradient IPNs [7, 76]).

It is also highly probable that curing of oligomer blends leads to networks with a higher content of ring fragments than when individual components are subjected to homopolymerization; moreover, the poorer the thermodynamic affinity, the higher the content of ring structures [12, 77].

It was found that during the cure of the blends of diphenylolpropane diglycidylic esters with dichlorodiaminophenylmethane, the maximum attainable concentration of "highly bonded" fragments (junctions with branching functionalities equal to 3 and 4) and the largest lengths of the chains leaving these junctions are obtained at a stoichiometric ratio between the components [78].

In the studies of curing of the blends of urethane-forming oligomers [12, 79], it was unambiguously demonstrated that the chemical junctions control the number and distribution of the physical junctions.

Finally, theoretical treatment of the effects of MMD and FTD of an oligomer on the topological structure of the three-dimensional polymer, which was reported in [79], may be extrapolated to oligomer blends by considering polydisperse oligomers as a mixture of oligomers of identical nature, that is, as oligomers that are cured according to the same mechanism but whose functionalities and lengths of oligomeric block are different. As a result of such theoretical analysis, quantitative relationships were obtained for the network density distribution, ratios of the fraction of pendant chains to that of the chains between the junctions, and limiting conversions as functions of the relative contents of zero-, mono-, bi-, tri-, ..., functional oligomers.

Although the disclosed regularities were modeled and verified for linear and cross-linked polyurethanes based on oligodiols, triols or tetraols and diisocyanates in the presence of monofunctional additives, Entelis et al. [79] believe that these regularities may be used to predict the topological structure of different classes of reactive oligomers.

Note that topological analysis of the structure of cured oligomer blends is a rather complicated problem, which has attracted appropriate attention only recently.

Supermolecular level. At this level, structural organization is controlled by intermolecular interaction between different components and the degree of order in the arrangement of structural elements.

Formation of a chemical bond between the chains must lead to a more compact packing because the covalent bond is much shorter than the van der Waals bond, provided that other conditions are the same.* However, the overall (i.e., over the entire volume of the sample) compaction is possible only when the chemical interaction between the different (and even the same) components does not violate the arrangement of other chain fragments with rather high packing perfection. However, this is not always the case. A chemical event may lead either to a more compact packing, which results from bond shortening, or cause disordering of the chains between the junctions. As a result of this competition, the overall contribution from these opposite tendencies to the variation of the free volume of the system may be either positive or negative.

Both cases were observed in experimental studies. The final outcome of this competition largely depends on the length of oligomer blocks. Flexible blocks acquire a regular packing more easily, provided that they are not too long and do not form coils. In that case, the system may be analyzed within the framework of the model of Flory's statistical coils with crossover. The final result also depends on the degree of cross-linking (the higher the degree of cross-linking, the poorer the conformational set available), on the testing temperature (whether it is above T_g or below it; in the glassy state the relaxation is retarded), etc. Within the framework of an alternative scheme [5], supermolecular organization of cured oligomer blends was qualitatively described in terms of "physical" and "chemical" clusters.

The packing density in cured oligomer blends is controlled by the molecular dimensions of the formed components. Cavities, voids, and other defects of one network may be filled by linear and cross-linked structural elements of the other component only when these elements fit into those defects. This feature is used to explain the differences in the properties of the IPNs prepared by simultaneous and consecutive curing. In the last case, small molecules of oligomer fill the accessible voids during the swelling and are then cured within those voids. In the case of simultaneous IPNs, different fragments grow in the course of the process simultaneously, although with different rates. Therefore, many defects of the network structure remain "unhealed" because of the different geometric dimensions of the growing fragments.

The packing density of the cured systems also depends on the degree of orientation of the molecules of the constituent components

---

* The shrinkage that accompanies the molding of thermosetting polymers is a manifestation of this effect.

in local supermolecular heterogeneities that were formed in the liquid phase. Formation of anisotropic structures in oligomer blends may occur both spontaneously [87] and by subjecting a system to different force fields. This statement is confirmed by the results reported in [82] where birefringence, IR dichroism, and X-ray diffraction were used to examine the structure of films of the blends of high-molecular-mass polystyrene ($M > 10^6$, $M_w/M_n = 1.1$) with oligooxyethylene dimethacrylates (DMOE-n, $n = 1, 3, 13$) before and after the curing with a beam of accelerated electrons (0–200 kGy). The films were drawn to $\lambda = 0$–1000%.

Table 3.3 lists the data on molecular orientation Δn, evaluated as described in [83], for some of the oligomer blends examined in [82] prior to curing (numerator) and after the curing by irradiation (denominator). Irradiation of the initial polystyrene (in the absence of oligomer) both in oriented and unoriented states did not affect Δn [82]. Therefore, uniaxile drawing of the oligomer blend (to $\lambda \leq$ 600–800%) increases the molecular orientation (by 1.5–3 times) of the oligomer only. The acquired orientation is then fixed (and even increases) in a solid phase obtained by radiation curing of the liquid blend. This unexpected result was confirmed by the X-ray diffraction and IR dichroism measurements.

Although the explanations suggested in [82] cannot be considered adequate, the fact that mechanical field causes the molecular reorientation in the associates (formation of cybotaxises) was convincingly demonstrated. This fact is very important because it opens up new prospects in controlling the supermolecular structure of oligomer blends by using rather simple technological means. It is essential that the anisotropic structures formed as a result of drawing are not destroyed or deteriorated by irradiation and concomitant chemical curing. Moreover, in some cases, the degree of order even increases, as is revealed by enhanced mechanical properties (see Section 4.3). Note also that an increase in the degree of orientation was also reported for polar systems subjected to preliminary orientation in an electric field and then polymerized by UV irradiation [86].

Morphological picture of the supermolecular structure of the cured composites is primarily controlled by the degree of cross-linking. For lightly cross-linked networks, all morphological forms known for amorphous and crystalline polymers (viz., globules, spherulites, lamellae, firbrils and others) were observed in experiment. Moreover, by acting as temporary plasticizers, the oligomers added in small amounts ($< 10\%$) to linear polymers may make the arrangement of macromolecules into regular structures easier. Subsequent curing fixes these regular structures.

Table 3.3. Variation of the molecular orientation in polymer–oligomer blends associated with uniaxial drawing and irradiation [82]

Blend	$\lambda, \%$	Dose, kGy	$-\Delta n \cdot 10^3$
PS + 50% DMOE-1	200	0/90	3.1/ 4.3
	400		7.2/ 11.4
	600		10.4/ 16.8
	800		30.5/ 44.8
PS + 20% DMOE-3	200	0/38	4.1/ 3.0
	400		9.8/ 11.0
	600		24.9/ 38.6
	800		50.2/ 52.2
	1000		36.1/ 54.6
PS + 50% DMOE-3	200	0/62	5.8/ 11.0
	400		8.7/ 30.4
	600		13.5/ 32.2
	800		26.3/ 64.4
	1000		42.9/149.8
PS + 80% DMOE-3	200	0/100	0.9/ 1.7
	400		3.2/ 4.5
	600		3.1/ 4.9
	800		3.9/ 5.6
	1000		6.9/ 21.6
PS + 50% DMOE-13	200	0/51	1.6/ 1.1
	400		4.4/ 5.3
	600		9.8/ 14.5
	800		15.8/ 22.7

As the concentration of network junctions increases, formation of perfectly packed morphological structures becomes more and more difficult. Chemical junctions disturb the short-range and long-range order in solid systems. For densely cross-linked systems ($> 10^{21}$ junctions/cm^3), only globules were observed in experiment. Only the size of the globules and their arrangement could be controlled by varying the components, their relative contents, and the curing conditions. **Colloidal level.** Cured oligomer blends are always heterogeneous. It is important to specify the dimensional scale that one associates with heterogeneity. Omitting the argumentation of different authors (this was scrutinized by Lipatov [7]), let us identify the primary element of structural heterogeneity in a cured oligomer blend with the phase structures that are nonequilibrium with respect to composition and the degree of separation between them. These are nanoparticles, as we defined them, or phasons, according to the terminology assumed in [24]. In terms of morphological notation, these primary elements are identified with the globules and nuclei [22]. Such structures are formed as a result of curing in single-phase oligomer blends or in each of the coexisting phases of the initially two-phase systems. Owing to "chemical quenching" (terminology of [5, 7]), these structures are kinetically stable.

Variation of the composition of nanoparticles can be described by a sinusoidal curve (Figs. 3.18 and 3.21); in extreme cases, this distribution may be either saw-tooth-shaped or sigmoidal. Naturally, the structure of cured oligomer blends (e.g., IPNs) may be represented as comprising two microscopic regions with an excess and deficiency of one of the components and a transition region between them [7].

The size of nanoparticles is controlled by the conditions of their formation, other conditions being the same. According to the data of electron microscopy and X-ray diffraction the linear dimensions of nanoparticles in different oligomer blends vary from several tens to several thousands of angstroms [29]. The degree of segregation of the components in the particles varies from 1 to 40% [24]. The size distribution for nanoparticles in the cured single-phase oligomer blends is usually monomodal (Fig. 3.22), although, under certain conditions, a more complex distribution function may be observed (see Table 3.2).

The curing scheme discussed in the preceding section (Fig. 3.18) suggests the existence of a densely cross-linked core, a lightly cross-linked structure "plasticized" at the outskirts of a particle by the second component, and interparticle (interglobular) matter. However, the concepts of transition layer and interfacial region have not been assigned strict physical meaning. Nevertheless, these regions must really exist and may be represented as structures whose composition is close to the average over the entire system.*

Extension of the transition layer and the volume fraction associated with this layer depend on the degree of segregation of the blend the components. According to the calculation resting on the relaxation data [36] for the cured blend PVC + TGM-3, the volume fraction of interfacial region varies from 10 to 40%. This variation depends

---

* Different models of transition layers in polymer blends (e.g., linear gradient model, sigmoidal model, equilibrium model, and their modifications) were discussed in detail in [23, 28, 80].

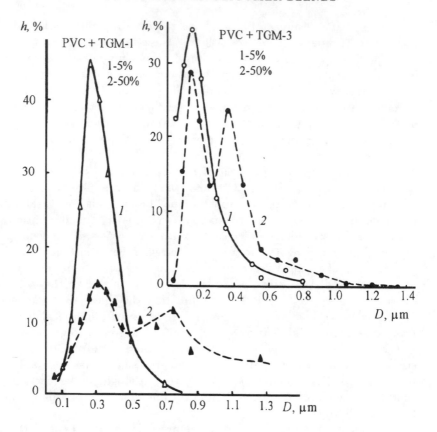

Figure 3.22. Particle size distribution in cured blends of PVC with (*a*) oxyethylene dimethacrylate and (*b*) trioxyethylene dimethacrylate. Oligomer content: (*1*) 5, (*2*) 50 wt %. Data of electron microscopy.

on the relative contents of the components and the curing conditions. According to X-ray diffraction studies, the volume fractions of transition layers in IPNs based on polyurethane and polyurethane acrylate vary in the same interval [72].

Different experimental methods give different estimates for the thickness of interfacial layers. According to X-ray diffraction measurements, the upper limit for the thickness of interfacial layer is 5–10 nm [77].

For a process engineer, it is very important to know the nature of a dispersion medium. Indeed, it is the dispersion medium that is mostly responsible for the macroscopic properties of disperse systems.* For cured pure oligomers, the situation is illustrated by the scheme in Fig. 3.6. Continuous phase is formed by the stuck particles of a densely cross-linked network.

In the case of oligomer blends the situation is more complicated. The structure of continuous phase is primarily controlled by the ratio between the contents of the components in initial mixture. At low contents of the second component ($\leq 5\%$), its amount may be insufficient to reorganize the structure of the continuous phase. In that case, this structure is controlled by the nature of the component that is in excess.

As the concentration of the second component is increased within the limits that allow the single-phase state of the initial system to be retained, the amount of matrix-forming component that is bonded during the cure with the second component increases. For example, for a blend of PVC (70%) with oligoester acrylate, curing was reported to result in a twofold decrease in the amount of free PVC [36]. The structure of dispersion medium in those composites is heterogeneous in composition and may be roughly identified as a "solid solution" enriched with one of the components of the blend.

In the case of symmetric composition of the solution of the initially single-phase oligomer blend, curing may lead to a structure characterized by double phase continuity (according to [5]) or to the so-called "inversion structures" (according to [23]). In that case, each of the phases may be considered macroscopically continuous. However, on a microscopic level both phases are heterogeneous.

Curing of oligomer blends that were initially in the two-phase state leads to an even more complicated picture. This picture is outlined in Fig. 3.20. Micron-size particles are dispersed in a continuous phase composed of compactly packed nanoparticles, which in their turn are dispersed in a dispersion medium, which is a "solid solution" enriched with one of the components of the blend. The structure formed is in fact a "phase-in-phase-in-phase" structure described in [60]. Heterogeneity of both the continuous and the disperse phases

---

* Introducing disperse particles into a matrix may help in realization of the important properties potentially inherent to the dispersion medium that by some reason are not observed when the matrix does not contain the dispersed particles. However, incorporation of the second component may also deteriorate the structure characteristic of the pure polymer.

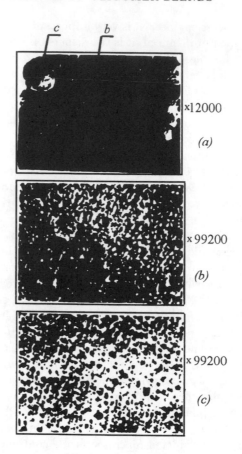

Figure 3.23. Electron micrographs: (*a*) cured blend of SKI-3 isoprene rubber and 10% α-trimethacryl-ω-methacryl-pentaerythrytol-di(dimethacryl-pentaerythrytol-adipinate) (magnification ×12 · 10³); (*b*) dispersion medium; (*c*) dispersed phase (magnification ×99 · 10³).

was confirmed by electron microscopy (Fig. 3.23) and the bimodal particle size distribution observed in the initially two-phase systems (Fig. 3.22).

REFERENCES

1. Berlin, A.A., Kefeli, T.Ya., and Korolev, G.V., *Polyefirakrilaty* (Polyester Acrylates), Moscow: Nauka, 1967 (in Russian).
2. Berlin, A. A. and Mezhikovskii, S.M., *Zh. Vses. Khim. o-va im. D.I. Mendeleeva,* 1975, no. 5, p. 531.
3. Kercha, Yu.Yu., *Fizicheskaya khimiya poliurethanov* (Physical Chemistry of Polyurethanes), Kiev: Naukova Dumka, 1975 (in Russian).
4. Bucknall, C.B., *Toughened Plastics,* London: Pergamon, 1977.
5. Sperling, L.H., *Interpenetrating Polymer Networks and Related Materials,* New York: Plenum, 1981.
6. Dontsov, A.A., Kanauzova, A.A., and Litvinova, T.V., *Kauchuk-oligomernye kompozitsii* (Rubber–Oligomer Composites), Moscow: Khimiya, 1986 (in Russian).
7. Lipatov, Yu.S., *Fiziko-khimicheskie osnovy napolneniya polimerov* (Physico-Chemical Foundations of Polymer Filling), Moscow: Khimiya, 1991 (in Russian).
8. Sedov, L.N. and Mikhailova, Z.V., *Nenasyshchennye poliefiry* (Unsaturated Polyesters), Moscow: Khimiya, 1977 (in Russian).
9. Bagdasar'yan, Kh.S., *Teoriya radikal'noi polimerizatsii* (Theory of Radical Polymerization), Moscow: Nauka, 1966 (in Russian).
10. Korshak, V.V. and Vinogradova, S.V., *Ravnovesnaya polikondensatsiya* (Equilibrium Condensation Polymerization), Moscow: Nauka, 1968 (in Russian).
11. Korshak, V.V. and Vinogradova, S.V., *Neravnovesnaya polikondensatsiya* (Non- Equilibrium Condensation Polymerization), Moscow: Nauka, 1972 (in Russian).
12. Irzhak, V.I., Rozenberg, B.A., and Enikolopov, N.S., *Setchatye polimery* (Polymer Networks), Moscow: Nauka, 1979 (in Russian).
13. Berlin, A.A., Korolev, G.V., Kefeli, T.Ya, and Sivergin, Yu.M., *Akrilovye oligomery i materialy na ikh osnove* (Acrylic Oligomers and Related Materials), Moscow: Khimiya, 1983 (in Russian).

14. Mezhikovskii, S.M., *I Vses. konf. po khimii i fizikokhimii polimer-izatsionnosposobnykh oligomerov* (First All-Union Conf. on Physical Chemistry of Polymerizable Oligomers), Chernogolovka: Akad Nauk SSSR, 1977, vol. 2, p. 362 (in Russian).

15. Nadzharyan, S.N., *Bitumno–oligomernye kompozitsii dlya sozdaniya materialov stroitel'nogo naznacheniya* (Bituminous–Oligomer Blends for Construction Engineering Materials), Cand. Sc. (Technical Sc.) Dissertation, Moscow: 1991 (in Russian).

16. Zapadinskii, B.I., *Sovremennye napravleniya sinteza termostoikikh polimerov oligomernym metodom* (Modern Trends in the Synthesis of Thermally Stable Polymers from Oligomers), Chernogolovka: Akad. Nauk SSSR, 1990 (in Russian).

17. Lipatova, T.E., *Kataliticheskaya polimerizatsiya oligomerov i formirovanie polimernykh setok* (Catalytic Polymerization of Oligomers and Formation of Polymer Networks), Kiev: Naukova Dumka, 1974 (in Russian).

18. Berlin, A.A. and Shutov, F.A., *Penopolimery na osnove reaktsionnosposobnykh oligomerov* (Foamed Polymers Based on Reactive Oligomers), Moscow: Khimiya, 1978 (in Russian).

19. Siling, M.I., *Polikondensatsiya: Fiziko-khimicheskie osnovy i matematicheskoe modelirovanie* (Condesation Polymerization: Physico-Chemical Principles and Mathematial Modeling), Moscow: Khimiya, 1988 (in Russian).

20. Dusek, K., In: *Kompozitnye polimernye materialy* (Polymer Composites), Kiev: Naukova Dumka, 1975 (in Russian).

21. Rozenberg, B.A., *Problemy fazoobrazovaniya v oligomer–oligomernykh sistemakh* (Phase Formation in Oligomer–Oligomer Systems), Chernogolovka: Akad. Nauk SSSR, 1986 (in Russian).

22. Korolev, G.V., *Mikrogeterogennyi mekhanism trekhmernoi radikal'noinitsiirovannoi polimerizatsii* (Microheterogeneous Mechanism of Three-Dimensional Radical Polymerization), Chernogolvka: Akad. Nauk SSSR, 1986 (in Russian).

23. Lipatov, Yu.S., *Kolloidnaya khimiya polimerov* (Colloid Chemistry of Polymers), Kiev: Naukova Dumka, 1984 (in Russian).

24. Shilov, V.V. and Lipatov, Yu.S., In: *Fizikokhimiya mnogokomponentnykh polimernykh sistem* (Physical Chemistry of Multicomponent Polymer Systems), Kiev: Naukova Dumka, 1986, vol. 2, p. 25 (in Russian).

25. Olabisi, O., Robeson, L., and Show, M., *Polymer Miscibility,* New York: Academic, 1979.

26. Rao, C.N. and Rao, K.J., *Phase Transitions in Solids,* McGraw Hill, 1978.

27. Cahn, J.W., *Trans. Met. Soc. AIME,* 1968, vol. 242, no. 2, p. 166.

28. Kuleznev, V.N., *Smesi polimerov* (Polymer Blends), Moscow: Khimiya, 1980 (in Russian).

29. Mezhikovskii, S.M., *Struktura i svoistva polimer–oligomernykh sistem i kompozitov na ikh osnove* (Structure and Properties of Polymer–Oligomer Systems and the Derived Composites), Doctoral (Technical Sc.) Dissertation, Moscow, 1983.

30. Zhil'tsova, L.A., Mezhikovskii, S.M., and Chalykh, A.E., *Vysokomol. Soedin., Ser. A,* 1985, vol. 27, no. 3, p. 587.

31. Mezhikovskii, S.M., Zhil'tsova, L.A., and Chalykh, A.E., *Vysokomol. Soedin., Ser. B,* 1986, vol. 28, no. 1, p. 42.
32. Mezhikovskii, S.M., Chalykh, A.E., and Zhil'tsova, L.A., *Vysokomol. Soedin., Ser. B,* 1986, vol. 28, no. 1, p.
33. Berlin, A.A., *I Vses. konf. po khimii i fizikokhimii polimerizatsion-nosposobnykh oligomerov,* (Abstracts of Papers, First All-Union Conf. on Physical Chemistry of Polymerizable Oligomers), Chernogolovka: Akad. Nauk SSSR, 1977, vol. 1, p. 8 (in Russian).
34. Mezhikovskii, S.M., *Nekotorye problemy fizikokhimii polimer–oligomer-nykh sistem i kompozitov na ikh osnove* (Some Problems in Physical Chemistry of Polymer–Oligomer Systems and the Related Composites), Chernogolovka: Akad. Nauk SSSR, 1986 (in Russian).
35. Mezhikovskii, S.M., *Vysokomol. Soedin., Ser. A,* 1987, vol. 29, no. 8, p. 1571.
36. Kotova, A.V., *Fazovaya struktura polivinilkhloridnykh–oligoefirakrilat-nykh sistem i kompozitov na ikh osnove. Termodinamicheskie i kinetich-eskie zakonomernosti formirovaniya* (Phase Structure of Polyvinyl Chloride–Oligoester Acrylate Systems and the Related Composites. Thermodynamic and Kinetic Regularities), Cand. Sc. (Chemistry) Dissertation, Moscow, 1988 (in Russian).
37. Volkov, V.P. et al., *Usp. Khim.,* 1982, vol. 51, no. 10, p. 1733.
38. Roshchupkin, V.P., Ozerkovskii, B.V., and Karapetyan, Z.A., *Vysokomol. Soedin., Ser. A,* 1977, vol. 19, no. 10, p. 2239.
39. Korolev, G.V., *I Vses. konf. po khimii i fizikokhimii polimerizatsion-nosposobnykh oligomerov* (Abstracts of Papers, First All-Union Conf. on Physical Chemistry of Polymerizable Oligomers), Chernogolovka: Akad. Nauk SSSR, 1977, vol. I, p. 144 (in Russian).
40. Mandelkern, L., *Crystallization of Polymers,* New York: McGraw-Hill, 1964.
41. Rozenberg, B.A., In: *Polimery-90* (Polymers-90), Chernogolovka: Akad. Nauk SSSR, 1991, vol. 1, p. 56 (in Russian).
42. Rozenberg, B.A., Nikitin, O.V., and Mezhikovskii, S.M., *II Vses. konf. "Smesi polimerov",* (Abstracts of Papers, Second All-Union Conf. on Polymer Blends), Kazan: Akad. Nauk SSSR, 1990, p. 8 (in Russian).
43. Wunderlich, B., *Macromolecular Physics,* New York: Academic, 1973, vol. 1.
44. Skripov, V.P. and Koverda, V.P., *Spontannaya kristallizatsiya pereokh-lazhdennykh zhidkostei* (Spontaneous Crystallization of Supercooled Liquids), Moscow: Nauka, 1984 (in Russian).
45. De Gennes, P., *Scaling Concepts in Polymer Physics,* Ithaca: Cornell Univ., 1979.
46. Shilov, V.V., *Formirovanie i osobennosti geterogennoi struktury mno-gokomponentnykh sistem* (Formation and Specific Features of Hetero-geneous Structure of Multicomponent Systems), Doctoral (Chemistry) Dissertation, Kiev, 1983 (in Russian).
47. Rowe, E. and Reiw, C., *Plast. Eng.,* 1975, vol. 31, no. 3, p. 45.
48. Griogor'eva, O.P., *Mikrofazovoe razdelenie pri formirovanii psevdovza-imopronikayushchikh setok* (Microphase Separation During theSynthe-sis of Pseudointerpenetrating Networks), Cand. Sc. (Chemistry) Dis-sertation, Kiev, 1984 (in Russian).

49. Chalykh, A.E., *Diffuziya v polimernykh sistemakh* (Diffusion in Polymer systems), Moscow: Khimiya, 1987 (in Russian).
50. Weibel, M. *et al., Polym. Adv. Technol.,* 1991, vol. 2, no. 2, p. 75.
51. Novikova, L.A., Dogadkin, B.A., and Tarasova, Z.N., *Vysokomol. Soedin., Ser. A,* 1968, vol. 10, no. 1, p. 211.
52. Avrasin, Ya.D. and Prigoreva, V.A., *Plast. Massy,* 1960, no. 1, p. 13.
53. Berlin, A.A., Kaplunov, I.Ya., and Barminov, V.A., *Plast. Massy,* 1966, no. 3, p. 5.
54. Cowepethwaite, G.F., SPE J., 1973, vol. 29, no. 2, p. 56.
55. Belozerov, N.V., *Struktura i svoistva vulkanizatov butadien–nitril'nykh kauchukov s nenasyshchennymi oligoefirami* (Structure and Properties of Vulcanizates of Butadiene–Acrylonitrile Rubbers with Unsaturated Oligoesters), Cand. Sci. (Technical) Dissertation, Yaroslavl', 1970 (in Russian).
56. Mash'yanova, I.M., *Dipol'naya relaksatstiya v oligomerakh so spetsificheskim mezhmolekulyarnym vzaimodeistviem i kauchuk–oligomernykh kompozitsiyakh na ikh osnove* (Dipole Relaxation in Oligomers with Specific Intermolecular Interaction and Related Rubber–Oligomer Composites), Cand. Sci. (Technical) Dissertation, Leningrad, 1990 (in Russian).
57. Kuz'minskii, A.S., Berlin, A.A., and Arkina, S.N., In: *Uspekhi khimii i fiziki polimerov* (Advances in Polymer Chemistry and Physics), Moscow: Khimiya, 1973 (in Russian).
58. Salmon, W.A. and Loan, L.D., *J. Appl. Polym. Sci.,* 1972, vol. 16, no. 3, p. 671.
59. Dontsov, A.A., *Protsessy strukturirovaniya elastomerov* (Structuration in Elastomers), Moscow: Khmiya, 1978 (in Russian).
60. Manson, J.A. and Sperling, L.H., *Polymer Blends and Composites,* New York: Plenum, 1976.
61. Bair, H.E. *et al., Macromolecules,* 1972, vol. 5, no. 2, p.114.
62. Lomonosova, N.V., Faizi, N.Kh., and Chikin, Yu.A., *Plast. Massy,* 1984, no. 12, p. 17.
63. Zadontsev, B.G., Yaroshevskii, S.A., Mezhikovskii, S.M., *et al., Plast. Massy,* 1984, no. 5, p. 9.
64. Zadontsev, B.G., Zapadinskii, B.I., and Mezhikovskii, S.M., *Plast. Massy,* 1984, no. 5, p. 18.
65. Yaroshevskii, S.A., *Polimer–oligomernye kompozitsionnye materialy na osnove lineinykh polimerov i oligoefirakrilatov* (Polymer–Oligomer Materials Based on Linear Polymers and Oligoester Acrylates), Cand. Sci. (Technical) Dissertation, Moscow, 1986 (in Russian).
66. Makhlis, F.A., *Radiatsionnaya khimiya elastomerov* (Radiation Chemistry of Elastomers), Moscow: Atomizdat, 1976 (in Russian).
67. Kuz'minskii, A.S., Kavun, S.M., and Kirpichev, V.P., *Fiziko-khimicheskie osnovy polucheniya, pererabotki i primeneniya elastomerov* (Physico-Chemical Principles of Production, Processing, and Application of Elastomers), Moscow: Khimiya, 1976 (in Russian).
68. Mal'chevskaya, T.D., *Formirovanie i svoistva vulkanizatov na osnove kauchuk– oligomernykh kompozitsii* (Formation and Properties of Vulcanizates Based on Rubber–Oligomer Composites), Cand. Sci. (Chemistry) Dissertation, Moscow, 1980 (in Russian).

69. Frenkel', R.Sh., *Issledovaniya v oblasti modifikatsii butadien–nitril'nykh kauchukov s tsel'yu polucheniya rezin povyshennoi rabotosposobnosti* (A Study on Modification of Butadiene–Acrylonitrile Rubbers with the Aim of Preparing Rubbers with Enhanced Performance), Doctoral (Technical) Dissertation, Moscow, 1978 (in Russian).

70. Lipatov, Yu.S., Mezhikovskii, S.M., Shilov, V.V., et al., *Kompoz. Polim. Mater.*, 1986, no. 28, p. 31.

71. Roginskaya, G.F., *Termodinamicheskie i kineticheskie zakonomernosti formirovaniya fazovoi struktury epoxi–kauchukovykh kompozitsii* (Thermodynamic and Kinetic Regularities of the Formation of Phase Structure of Epoxide–Rubber Composites), Cand. Sci. (Chemistry) Dissertation, Moscow, 1983 (in Russian).

72. Lipatov, Yu.S., Shilov, V.V., Gomza, Yu.P., and Kruglyak, N.E., *Rentgenograficheskie metody izucheniya polimernykh sistem* (X-Ray Methods for Studying Polymer Systems), Kiev: Naukova Dumka, 1982 (in Russian).

73. Rebrov, A.V., *Osobennosti strukturoobrazovaniya v sistemakh kauchuk–oligoefirakrilaty* (Specific Features of Structure Formation in Rubber-Oligoester Acrylate Systems), Cand. Sci. (Chemistry) Dissertation, Moscow, 1983 (in Russian).

74. Glansdorff, P. and Prigogine, I., *Thermodynamic Theory of Structure, Stability ,and Fluctuations,* London: Wiley, 1971.

75. Polak, L.S. and Mikhailov, A.S., *Samoorganizatsiya v neravnovesnykh fizikokhimicheskikh sistemakh* (Self-Organization in Non-Euilibrium Physico-Chemical Systems), Moscow: Nauka, 1983 (in Russian).

76. Sergeeva, L.M. and Lipatov, Yu.S., In: *Fizikokhimiya mnogokomponentnykh polimernykh sistem* (Physical Chemistry of Multicomponent Polymer Systems), Kiev: Naukova Dumka, 1986, vol. 2, p. 137 (in Russian).

77. Shilov, V.V., Lipatov, Yu.S., and Tsukruk, V.V., In: *Fizikokhimiya mnogokomponentnykh polimernykh sistem* (Physical Chemistry of Multicomponent Polymer Systems), Kiev: Naukova Dumka, 1986, vol. 2, p. 101 (in Russian).

78. Lantsov, V.M., *Strukturno-kineticheskaya neodnorodnost' molekul v oligomernykh, polimer–oligomernykh i polimernykh setkakh na ikh osnove* (Structural and Kinetic Nonuniformity of Molecules in the Derived Oligomer, Polymer–Oligomer, and Polymer Networks), Doctoral (Chemistry) Dissertation, Moscow, 1989 (in Russian).

79. Entelis, S.G., Evreinov, V.V., and Kuzaev, A.I., *Reaktsionnosposobnye oligomery* (Reactive Oligomers), Moscow: Khimiya, 1985 (in Russian).

80. Lebedev, E.V., In: *Fizikokhimiya mnogokomponentnykh polimernykh sistem* (Physical Chemistry of Multicomponent Polymer Systems), Kiev: Naukova Dumka, 1986, vol. 2, p. 74 (in Russian).

81. Andrianov, K.A. and Emel'yanov, V.N., *Usp. Khim.*, 1976, no. 10, p. 1817.

82. Lomonosova, N.V., *Vysokomol. Soedin., Ser. A*, 1992, vol. 34, no. 6, p. 48. (English translation: Polymer Science, 1992, vol. 34, no. 6, p. 492).

83. Stein, R., in: *Newer Methods of Polymer Characterization, Ke. B, Ed.*, New York, 1964.

84. Irzhak, V.I. and Rozenberg, B.A., *IV Vses. konf. po khimii i fizikokhimii oligomerov* (Abstracts of Papers, Fourth All-Union Conf. on the Chemistry and Physical Chemistry of Oligomers), Chernogolovka: Akad. Nauk SSSR, 1990, p. 23 (in Russian).
85. Irzhak, V.I., *V Vses. konf. po khimii i fizikokhimii oligomerov* (Abstracts of Papers, Fourth All-Union Conf. on the Chemistry and Physical Chemistry of Oligomers), Chernogolovka: Ross. Akad. Nauk, 1994, p. 20 (in Russian).
86. Broer, D.J., Gossink, R.G., and Hikmet, R.A., *Angew. Makromol. Chem.*, 1990, vol. 183B, p. 45.
87. Mezhikovskii, S.M., *Kinetika i termodinamika protsessov samoorganizatsii v oligomernykh smesevykh sistemakh* (Kinetics and Thermodynamics of Self-Organization in Blended Oligomeric Systems), Chernogolovka: Ross. Akad. Nauk, 1994 (in Russian).
88. Irzhak, V.I., Tai, M.L., Peregudov, N.I., and Irzhak, T.F., *Colloid. Polym. Sci.*, 1994, vol. 272, p. 523.
89. Irzhak, T.F., Peregudov, N.I., Irzhak, V.I., and Rozenberg, B.A., *Vysokomol. Soedin., Ser. B*, 1993, vol. 35, no. 7, p. 905; ibid. 1993, vol. 35, no. 9, p. 1545. (English translations: Polymer Science, Ser. B, 1993, vol. 35, no. 7, p. 993; ibid. 1993, vol. 35, no. 9, p. 1290.)
90. Irzhak, T.F., Peregudov, N.I., Tai, M.L., and Irzhak, V.I., *Vysokomol. Soedin., Ser. A*, 1994, vol. 36, no. 6, p. 914. (English translation: *Polymer Science, Ser. A*, 1994, vol. 36, no. 6, p. 754.)
91. Nikitin, O.V. and Rozenberg, B.A., *Vysokomol. Soedin., Ser. A*, 1992, vol. 34, no. 4, p. 139. (English translation: *Polymer Science*, 1992, vol. 34, no. 4, p. 365.)
92. Korolev, G.V. and Berezin, M.P., *Kineticheskie proyavleniya assotsiativnoi struktury zhidkikh oligomerov v protsessakh polimerizatsii i sopolimerizatsii* (Kinetic Manifestations of Associative Structure of Liquid Oligomers in Polymerization and Copolymerization), Chernogolovka: Ross. Akad. Nauk, 1994 (in Russian).

4 PHYSICOCHEMICAL ASPECTS OF THE MATERIALS SCIENCE OF OLIGOMER BLEND COMPOSITES

4.1. PRINCIPLES OF THE MORPHOLOGY CONTROL

The approaches to controlled structure formation in cured oligomer blends can be divided into two groups:*
(a) Purposeful action upon the initial mixture: control of the composition and temperature of the initial blend in the stages before curing.
(b) Purposeful modification of the curing process: variation of the curing regime for a given blend composition.

The effect of composition and technological parameters that can be simply varied in real technological schemes was analyzed elsewhere [1–3]. These parameters include (i) the initial oligomer content in the blend (c); (ii) the temperature of exposure (T_{exp}), at which the blend is stored during the time period between preparation and cure; (iii) the initiator concentration (I); (iv) the inhibitor concentration (In); and the curing temperature (T_{cure}).

Of the entire variety of structural parameters, we have selected two quantities that are comparatively readily determined in experiment and are responsible to a large extent for the mechanical properties of composites. These are the number of dispersed particles per unit volume (N) and their average size (R).

---

* Before reading this Section, it is recommended to turn back to paragraphs 2.2, 2.8, 3.8, and 3.9.

143

Table 4.1. The character of variation of the morphological parameters N and R with increasing values of the composition-technological parameters c, T_{exp}, I, In, and T_{cur}

Phase organization of the initial system	Morpho-logical parameter	Variation trend				
		c	T_{exp}	I	In	T_{cure}
Single-phase	N'_n	+	×	+	−	+
	R'_n	+	×	−	+	−
Two-phase (solution I)	N'_n	−	+	+	−	+
	R'_n	×	+	−	+	−
Two-phase (solution I)	N'_m	+	−	×	×	×
	R'_m	+	−	×	×	×
	N''_n	+	−	+	−	+
	R''_n	×	×	−	+	−

The mechanism of structure formation in the cured composites, which was considered in the preceding chapter, allows us to formulate the following rules for the controlled variation of morphological parameters of the final structure with in creasing values of c, T_{exp}, I, In, and T_{cure}. These rules are presented in the form of Table 4.1 and are suited for predicting the structure formation in the blends containing one nonreactive component, or the blends composed of two reactive components, of which one has a curing rate much higher than that of the other. As for the oligomer blend systems of all other types, only the dependence of morphology parameters on c and T_{exp} is valid among the trends given in Table 4.1, while the other correlations would require certain correction.

The notations in Table 4.1 are as follows: subscript 'n' refers to nanoparticles (or Ångström-size particles) of the first (prime) or second (double prime) type, and the subscript 'm' refers to the micronsize particles. Symbols '+','−', and '×' indicate that the given parameter of morphology increases, decreases, or remains unchanged with increasing value of a technological parameter of this column. The morphological changes upon curing are considered separately for the single-phase systems and the coexisting phases (solutions I and II) of two-phase oligomer blends. Data presented in Table 4.1 require some comments.

Effect of the Total Oligomer Content

In single-phase systems, an increase in c is equivalent to growing concentration of the reactive component. Therefore, provided all other conditions are equal, the N_n' and R_n' values increase with the total oligomer content.

A different situation is observed for the true two-phase systems. In this case, an increase in c has no effect on the oligomer concentrations in the coexisting phases, and leads only to a growth in the number and size of segregated parcels of the disperse phase (emulsion droplets). Upon curing, these are fixed in the system in the form of micron-size particles. Therefore, in the two-phase oligomer blend systems an increase in the total oligomer content gives rise to the N_m and R_m values.

As for the N_n' and N_n'' values, representing the number of nanoparticles formed in the disperse phase (it should be recalled that double-primed nanoparticles constitute the body of a micron-size particle), these would require a correction for the "scaling" factor. Indeed, if the function $N = f(c)$ is considered within the entire volume of the sample, then N_n'' must increase, while N_n' would accordingly decrease, with increasing c because the relative fraction of micron-size particles tends to increase. It is this very situation that is reflected in Table 4.1.

Calculation of the number of nanoparticles is quite simple. Since $N_n'/(N_n' + N_n'') + N_n''/(N_n' + N_n'') = 1$, the current N_n values vary with the total oligomer content c according to the relation $(N_n')_c = (N_n)_b(V_1/V_0)$, where $(N_n')_c$ and $(N_n')_b$ are the numbers of nanoparticles of the first kind for the given c value and for the equilibrium oligomer content in the dispersion medium at the binodal point, respectively, V_0 is the total volume of the system, and V_1 is the volume of the continuous phase.

As for the character of variation in the number of nanoparticles within a single micron-size particle, the N_n'' value is independent of c, that is, the concentration of oligomer in the disperse phase is independent of the total oligomer content. Therefore, the number of nanoparticles of the second kind, constituting the bodies of micron-sized particles, will be constant. This number can be calculated using the relation $N_n'' = (N_n'')_b(1 - V_1/V_0)$, where $(N_n'')_b$ is the number of nanoparticles of the second kind formed upon the curing of a solution with the given oligomer concentration at the corresponding binodal point on the phase diagram for the given temperature.

The scaling factor is also of importance for the analysis of the $R = f(c)$ function. Within each one of the coexisting phases, the

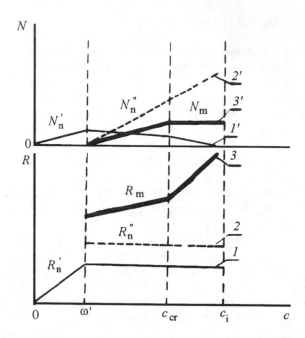

Figure 4.1. Variation of the particle number per unit volume (N) and their average size (R) as functions of the total oligomer content c. See the text for explanations.

R'_n and R''_n values are independent of c (see Table 4.1). However, taking into account a difference between absolute dimensions of the nanoparticles of the first and second kind $(R''_n > R'_n)$, the average particle size (irrespective of the type) normalized to the total particle number $(N'_n + N''_n)$ increases with the total oligomer content c. This is a very important circumstance, because experimental methods employed in the usual technological practice and research are uncapable to distinguish between the nanoparticles of various types, thus leading to incorrect conclusions (for more details, see [3]).

From the technological standpoint, it is highly important to know that the final composite morphology is predetermined to a considerable degree by the morphology developed in the blend at the beginning of curing. The laws of possible morphological changes in the initial systems were considered in Section 2.7. Here, we will only note that, if the viscosities of two components are different, an increase in the content of one (less viscous) component usually leads to violation of the morphological stabillity of the blend in the initial state. The value of c at which the oligomer blend system loses stability and begins to coalesce is denoted by c_c. This value usually corresponds to the oligomer content at which the coalescence time becomes comparable with the time of observation or the duration of a technological operation. Thus, the character of variation of the N_m and R_m values as functions of c may differ, depending on the particular composition domain to which the given blend belongs. For $c < c_c$, an increase in the oligomer content in the initial blend will lead to increasing number and size of the segregated parcels and, hence, to growing N_m and R_m values upon curing. This situation is reflected in Table 4.1.

For $c > c_c$, the coalescence leads to increasing dimensions of the segregated parcels of phase inclusions (emulsion droplets) at the expense of decreasing number of these formations. In the cured composite, this behavior will be manifested by different trends in the variatrion of R_m and N_m values (this case is not represented in the table).

Figure 4.1 illustrates the general trends in the variation of R and N values as functions of the oligomer content c for all types of particles formed during the cure of oligomer blend systems. These laws must be taken into account in using Table 4.1.

Effect of the Temperature of Exposure

In single-phase oligomer blend systems (we assume that a system has attained the equilibrium state) an increase in the temperature of exposure (at which the blend is stored during the time period between preparative and curing operations) has no effect on the morphological parameters of blend composites.* Of course, we assume that the temperature increase does not produce thermal initiation in the system, leading to the gel formation and premature curing of the blend during storage.

In the two-phase blends, the increase in T_{exp} always leads to increasing concentration of an oligomer in the dispersion medium (in systems possessing UCST, to which most of the oligomer blends belong). This, in turn, leads to an increase in N_n' and R_n' at the curing

---

* Here, we are speaking only of the morphological changes. It is suggested (since no experimental data are available) that T_{exp} influences the supermolecular organization of the liquid phase, in particular, the structural parameterers of sybotaxises. This may also affect the topological characteristics of blend composites.

temperature ($T_{cure} \geq T_{exp}$). Earlier [4] we presented arguments and experimental evidence for the fact that an increase in the temperature from T_{exp} to T_{cure} does not lead to changes in the phase organization of real systems, and the subsequent curing only fixes the morphology developed at $T = T_{exp}$.

In the absence of chemical reactions, an increase in T_{exp} results in reducing size and decreasing number of microparcels of the disperse phase. Upon curing, this is manifested by lower R_m and N_m values, and automatically leads to a decrease in N_n''.

Table 4.1 indicates that R_n'' is independent of T_{exp}, which reflects the fact that in many oligomer blends the equilibrium concentrations of the main (oligomer) component in the phase of solution II are small (the right-hand branch of the binodal approaches the temperature axis). However, in the case when the phase diagrams are symmetric and the solubility of oligomer in the second component is significant (information pertaining to particular systems can be found in the reference literature [5]), the increase of T_{exp} leads to decreasing or increasing oligomer concentrations in the disperse phase in systems with UCST or LCST, respectively. Therefore, the character of the variation of R_n'' in a particular system must somewhat change as compared to that presented in Table 4.1.

Another possible correction of the data in Table 4.1 is related to the fact that increasing T_{exp} in two-phase systems leads to decreasing viscosity of the dispersion medium and the correspondng shift of the c_c value. If the disperse phase exhibits coalescence at a given c_c value and a selected duration of the blend exposure at T_{exp}, the trend of R_m in Table 4.1 may change from plus to minus.

Effect of the Initiator Concentration

An increase in the I value is equivalent to growing number of the branching centers. During the cure, this must lead to increasing number of nanoparticles. This situation may take place both in the initially single-phase blends, and in each phase of the initially two-phase systems, as manifested by an increase in N_n' and N_n''. For $c = $ const, the increase in the number of particles would automatically result in decreasing size. The N_m and R_m values are independent of I, because a high rate of curing in solution II ensures fixation of the size of dispersed particles, established in the initial medium at the preparation stage, irrespective of the initiator concentration.

Effect of the Ihibitor Concentration

The character of morphological changes accompanying an increase in In is opposite to that induced by increasing I: the higher the In value, the lower is the number of branching centers, which eventually results in a lower number and greater dimensions of nanoparticles.

Variation of the content of inhibitor in an oligomer blend system does not affect the N_m and R_m values, because the I value is usually selected so as to ensure that the efficiency of initiator would exceed that of the inhibitor, and, as was shown above, the initiator has no effect on N_m and R_m. The inhibitor affects only the topology of micron-size particles, rather than their morphology.

Effect of the Temperature of Curing

With respect to the composite morphology, an increase in T_{cure} is analogous to a growth in the efficiency of initiation. Therefore, the change of T_{cure} will affect the N and R values of Ångström- and micron-size particles in the same way as does the variation of I considered above.

It should be emphasized that the qualitative character of results presented in Table 4.1 and in Fig. 4.1 is not affected by the nature of an oligomer blend system. Various blends may only differ, depending on their phase equilibria, by the particular intervals of c and T in which the general trends are valid.

Finally, it must be noted that conclusions derived from the structural-thermodynamic model describing the curing of oligomer blends (these conclusions serve as a basis of the pattern presented in Table 4.1), concerned only two morphological parameters—the number and dimensions of dispersed inclusions. Naturally, the other structural parameters (the width of an interphase boundary layer, the coefficient of molecular packing, the density of cross-linking, etc.) are connected to the phase equilibrium by relationships different from those known for the N and R quantities. However, this approach is suited for the search of correlations between an arbitrary structural parameter (at any level of the organization hierarchy) of cured composites and thermodynamic characteristics of the initial mixture. On this basis we can predict, using the phase diagrams of oligomer blend systems, the morphology of cured composites and control this morphology by simple technological methods.

4.2. RELATIONSHIP BETWEEN THE MORPHOLOGY OF SOLID COMPOSITES AND THEIR MECHANICAL PROPERTIES

Now we will consider the possibility to control the strength properties of cured oligomer blends by directed modification of their morphology parameters. To this end, we must first formulate the criteria by which it is possible to select the relevant structural parameters and determine their ranges providing the optimum mechanical properties of a given class of composites.

Requirements to the Morphology of Cured Oligomer Blends

The strength properties of heterogeneous materials are determined mostly by the deformation behavior of the continuous phase, because this component primarily accepts the load energy applied to the system. Behavior of the disperse phase depends on a number of factors and may either increase or decrease the strength of the system. We will not discuss here various theories of fracture and the details of various mechanisms involved in the strengthening of polymeric materials. These questions are thoroughly analyzed in the literature (see, e.g., [6–19]), and we will a priori accept the following basic concepts.

As is known, a build-up of stress σ during deformation of a polymeric material leads to the appearance of multiple submicro- and microcracks (crazes) originating at the structural defects. These cracks merge together and grow to form a macrocrack responsible for the breakage of material into parts (fracture). The problem of material strengthening reduces to either creating a defect-free structure or, since this is virtually impossible, preventing crazes from growth by "quenching" the fracture energy γ.

At least two mechanisms have been established by which heterogeneous inclusions (particles) may influence the process of fracture in polymers. These mechanisms are based on the brittle fracture or plastic shear flow of a part of the matrix between the particles. Which mechanism is operative—depends on the dimensions of dispersed particles.

If the average size of the particles obeys the condition $D \geq D_{\mathrm{cr}}$ (where D_{cr} is the critical size of the zone of plastic deformation at the crack apex), the growing crack would either become blunt upon collision with a particle, thus loosing all or a part of the energy, or has to by-pass the particle, also partly loosing the energy as result of increasing pathlength or branching. Evidently, the magnitude of breaking stress is proportional to the volume fraction c of the disperse phase.

If $D < D_{cr}$, then the shear mechanism becomes operative. In this case, the energy of fracture of the heterogeneous system is described by the formula

$$\gamma = \frac{8}{\pi} D_{cr} \cdot \frac{\sigma_y}{E}, \qquad (4.1)$$

where σ_y is the yield stress (induced rubberlike elasticity limit) and E is the elasticity modulus of the matrix. Because D_{cr} increases, and σ_y decreases, with increasing volume fraction of the disperse phase, the function $\gamma = f(c)$ exhibits an extremal (nonmonotonic) character.

Apparently, the maximum resistance of polymers to cracking would be observed if both mechanisms of interaction of the growing cracks with particles of the disperse phase were operative. Note that the former mechanism is most efficient when the system contains rather coarse particles, and the latter mechanism is favored by fine particles.

Since the probability of collisions of a growing crack with the particles is proportional to their concentration, we may expect in the first approximation that the greater the number of particles, the higher is the effect. However, not every particle may serve as a "trap" for the cracks and dissipate their energy. The "active" centers will be represented only by dispersed particles with dimensions comparable with (or slightly exceeding) the radius of curvature of the crack. Investigations of the deformation of polymers showed the appearance of cracks with a big scatter of the curvature radii, ranging from 20 to 2000 Å. Therefore, the maximum effect is expected in systems featuring a broad distribution of inclusions with respect to their dimensions. If the growing cracks meet no particles of appropriate dimensions, capable of dissipating the cracking energy, the probability of formation of a critical macrocrack increases.

There are also certain restrictions on the maximum particle size. One of the limitations is related to the fact that some large particles may act as sources of dangerous defects by themselves. For example, a crack nucleated at such a particle may immediately become critical. The upper limit of the particle dimensions, above which the strength of material sharply decreases, is designated by D_{lim}. It should be emphasized that here we are speaking of the maximum size, probably, of a single particle capable of inducing a catastrophic fracture of the material, rather than of the average parameter. The D_{lim} value is determined by the nature of the material and the conditions of deformation. According to the published data, the magnitude of D_{lim} may vary within at least three decimal orders. For example, from 100 μm in a filled polystyrene to 0.25 μm in an ABS-plastic-filled butadiene-styrene rubber.

Finally, we must take into account that a dispersed particle may serve as an obstacle for the crack only if the particle resides on the crack path. The maximum probability of collisions between cracks and particles is observed if the centers of particles are spaced by no more than 2.5D.

Naturally, the above considerations give by no means a complete list of the structural requirements necessary to ensure the optimum physico-mechanical properties of the material. We have analyzed on-ly the morphological criteria and did not consider some other factors such as (i) the condition of good adhesion (up to the chemical interac-tion) at the phase boundary, (ii) the importance of having close values of the thermal expansion coefficients of the matrix and inclusions (a significant difference may result in the polymer–particle phase separa-tion even before the stress application), (iii) the necessity of a signif-icant difference between the moduli of matrix and dispersed phases, etc.

Thus, the best strength properties of the composites are expected in the cured oligomer blend systems with uniformly dispersed hetero-geneous inclusions. The inclusions must have a broad distribution of particle dimensions, but the upper particle size is limited.

A question naturally arises as to how to create a structure satisfy-ing all the above requirements. An answer to this question is obtained by analysis of the data presented in Table 4-1.

Optimization of Breaking Strength

An engineer can meet a problem of how to increase the breaking (tensile) strength of an existing commercial material or create a new oligomer-blend-based material with increased σ_b value. We may well try to solve this task on a technological level, by selecting an optimum composition (ratio of the blend components) and finding appropriate process conditions.

An increase in the composite strength can be achieved by different approaches, depending on the phase organization of the initial blend.

Single-phase states. Increasing concentration (c) of a com-ponent responsible for the disperse phase formation upon cure leads to a growth in the number (N_n) and size (R_n) of disperse particles and, hence, to an increase in σ_b. This law is confirmed in most ex-periments, which show that σ_b is proportional to c. However, the strength of composites ontained upon cure of the single-phase composites never reaches a maximum σ_b level that can be expected for a given pair of blend components. This is caused by the fact that the condition of having a broad size distribution of the inclusions fails to be valid. The distribution width can be extended by properly setting the cure rate constants, which may even lead to a polymodal distribution function (see Section 3.7). However, this method of control always implies an increase in the initiation rate, which leads to a decrease in the average particle size and, hence, in σ_b. Therefore, the σ_b value of a cured single-phase oligomer blend is, even in the best case, a compromise between contributions of several factors acting in the opposite directions.

Two-phase states. This situation offers a markedly greater number of variants for acting upon a blend system so as to provide conditions ensuring the maximum growth in σ_b. Indeed, the cure of two-phase blends leads to the formation of structures containing a greater number of dispersed particles as compared to that in the single-phase state.

The two-phase systems are characterized by a greater width of the inclusion size distribution. Indeed, these system contain, besides the "nanoparticles" formed in the dispersion medium, also "micron" particles of the dispersed phase, which are composed of "nanoparticles" of the second type. By increasing c and selecting appropriate regimes of blending (see Section 2.7), we may provide a uniform distribution of the "micron" particles over the volume with a distance between them not exceeding $2.5\,D$ (D is the average particle diameter). However, another important condition is that the size of drops of the phase of solution II in the initial mixture (and, hence, the size of "micron" particles), which is fixed by curing, must not exceed $D_{\lim}$. This reqirement is illustrated in Fig. 4.2, showing variation in the morphology of a blend of polybutadiene rubber (SKD grade) and oligoester acrylate (MGF-9 grade), depending on the component ratio, and a curve of vulcanizate strength versus composition. In contrast to the initially single-phase systems, in which the increase in c is accompanied by a linear growth of σ_b [20–23], the σ_b versus c curve of a two-phase oligomer blend exhibits an extremum. The relative content of oligomer corresponding to the formation of critical drops with an average diameter of $D_{\lim}$ (and the maximum σ_b value) is designated by c_{crit}.

In order to increase σ_b, we may vary not only the blend composition, but the blend storage temperature regime as well. For example, it is possible to take an oligomer blend with a composition corresponding to the drop size definitely above the $D_{\lim}$ value (i.e., $c > c_{crit}$) and,

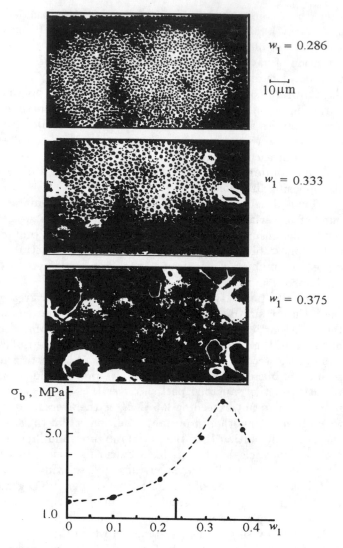

Figure 4.2. Morphology of uncured blends based on a *cis*-polu-butadiene rubber with various contents of α,ω-dimethacrylate-bis(trioxyethylene)phthalate oligomer. The curve shows a concentration dependence of the breaking strength of cured composites (with an arrow indicating the oligomer content corresponding to phase separation in the system).

hence, with the breaking strength occurring within descending branch of the σ_b versus c curve. Prior to the cure, the blend is exposed for some time to a temperature below the cure temperature but above the blending temperature.* During this exposure, the blend structure will change as follows. First, the oligomer concentration in the continuous phase increases and, second, the size of particles of the dispersed phase decreases. The exposure temperature (T_{exp}) and duration are selected so as to ensure that the drop size would decrease below D_{lim}. On the subsequent cure of this system, a composite will form with N_n smaller as compared to that of a composite treated at a lower T_{exp} and R_m satisfying the condition of Eq. (4.1). As a result, the breaking strength obviously increases.

There are some other methods that allow us to increase the range of variation of the N and R values for the controlled formation of composite structure. These approaches are not as obvious as the above direct conclusions from Table 4.1. Now we will consider two methods of this kind.

The first method is to modify an oligomer blend by adding surfactants, which are known to produce various effects on heterogeneous polymer systems [23–28]. This problem was partly considered in Section 2.7. Here, the idea is that introducing surfactants into a two-phase blend with a component ratio corresponding to $c > c_{crit}$ would lead to a decrease in the size of microparcels of the phase of solution II as a result of reduced surface tension. Subsequent cure of this blend would yield a structure with $R_m \leq D_{lim}$ and, since the c value is unchanged, the N_m would necessarily increase. Thus, a structure with a greater number of dispersed particles and an optimum distribution of their dimensions is created, while a high oligomer concentration in coexisting phases is retained. Evidently, the σ_b level in these systems must increase.

Other possible advantages of adding surfactants to oligomer blends consist in stabilization of the emulsion, increasing morphological stability, and extending the range of T_{exp} by preventing the coalescence at elevated temperatures. From technological standpoint, the latter circumstance is especially important at $c > c_{crit}$.

One feature of using surfactants for the modification of oligomer blends is worth of special mentioning. Prof. A.A. Berlin suggested that reactive monofunctional methacrylic esters of long-chain fatty acids can be used as efficient emulsifiers for oligomer systems. These

---

* The upper boundary of T_{exp} is limited by two conditions: (i) no chemical reactions must take place and (ii) coalescence must not be activated.

molecules are not only capable of emulsifying the system, but eventually (on curing) enter into chemical reactions with the blend components. This not only excludes many undesirable side effects introduced by nonreactive surfactants into disperse polymer systems (e.g., eliminate the Rebinder effect), but may open possibiities to change some other (besides the morphology) structural properties of composites. In this way we can simultaneously control both static and dynamic properties of the materials. These new possibilities are illustrated by Table 4.2 presenting the properties of elastomers based on a two-phase blend of an acrylate-butadiene rubber (SKN-18 grade) and an oligoester acrylate cured in the absence and presence of surfactants. This interesting direction in the oligomer blend modification is considered in more detail in [20, 22, 29].

Table 4.2. Physicomechanical characteristics of elastomers based on an acrylate-butadiene rubber (SKN-18 grade) [20].*

Parameter	Oligoester acrylate, wt. fract.					
	Without surfactant			With 6% surfactant**		
	0	30	50	0***	30	50
σ_b	2.0	6.0	10.0	2.0	9.0	15.0
l,%	500	450	350	500	450	350
z,%	10	12	14	12	12	14
Hardness (TM-2, rel. units)	30	43	57	31	50	63
Resistance to tear, kN/m	5	15	30	5	40	55
Dynamic durability for repeated compressions, 103 cycles	100	800	600	150	1000	800

Notes: *Blend composition (wt. fract.): SKN-18 (100), zinc oxide (5), peroximon (1), and an oligoester acrylate. The oligoester acrylate is a D-20/50 compound comprising a mixture of MDF-1 + 7-20 in a 1 : 1 ratio; vulcanization regime: $150°C \times 40$ min; surfactant:

$$CH_2 = \overset{\overset{\displaystyle CH_3}{|}}{C} - \overset{\overset{\displaystyle O}{\|}}{C} - O - (CH_2)_n - CH_3 , \quad n = 7-18.$$

**Calculated with respect to the oligoester acrylate.
***Surfactant content, 1.5 wt. fract.

Another efficient method for controlling the structure and properties of cured composites is to use a mixture (compound) of various oligomers as the oligomer component of the blend. The use of such compounds has a number of advantages, including the possibility to control viscosities of both the oligomer component and the blend, to improve the blending technology, to increase the variety of molecular structures, etc. Compounds used for the purpose are composed of oligomers with markedly differing solubilities in the second component of the blend (polymer, oligomer, or monomer). This method of structural control was studied for detail in polymer–oligomer systems [22, 30, 31].

The idea is essentially as follows. On adding a mixture of two oligomers (the well and poorly soluble ones) to a linear polymer, we may select their ratio such that one oligomer component would ensure a high oligomer concentration in the phase of solution I (on curing, this must eventually yield a large number of nanoparticles), while the other oligomer component would account for the formation of microparcels of the phase of solution II with dimensions close to D_{lim} (when fixed by curing, the latter would ensure the necessary size distribution of micron particles). Optimum dimensions of the inclusions are naturally different for each particular composite, but are readily controlled by the ratio of oligomers in the mixture and the total content of the oligomer compound in the blend. Information necessary for selecting the correct ratios is provided by phase diagrams. Let us consider a particular example.

A polybutadiene rubber (e.g., of the SKD grade) is modified by a compound based on oligoester acrylates (MDF-1 and 7-1). First, we use the phase diagrams (Fig. 4.3) to determine equilibrium room-temperature solubility concentrations ceq for these oligomers in the rubber. According to [32], these are 22–23 wt.% and 4–5 wt.% for MDF-1 and 7-1, respectively. By jointly considering the σ_b vs. c curves and the morphology data (see Fig. 4.2), we may select the c_{eq} values corresponding to the maximum or the beginning of descending branch for each oligomer. Because the use of the oligomer compound leads to saturation of the solution simultaneously with both oligomers, the c_{eq} value must be reduced in the necessary proportion. According to estimates, the formation of inclusions with dimensions $R_m = D_{lim}$ is achieved by using a compound with the concentrations of MDF-1 and 7-1 not exceeding 20 wt.% and 2.5–3.5 wt.%, respectively. This implies that the ratio of oligomers in an optimum compound, which would ensure attaining the best strength properties, must be taken equal to 5 : 1, rather that 1 : 1 or 1 : 2 typically used in commercial compositions. Note, however, that the commercial compositions

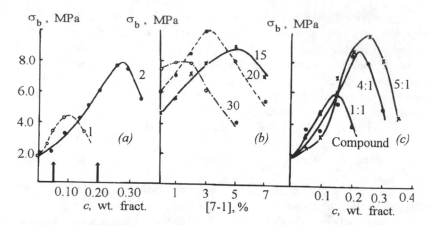

Figure 4.3. The plots of σ_b for cured blends of a polybutadiene rubber (SKD grade) with oligomer compounds of α-trimethacryl-ω-methacryl-pentaerythritol(dimethacryl-pentaerythritol-adipinate) (7–1) and α, ω-dimethacrylate-bis(dioxyethylene)phthalate (MDF-1):

(*a*) versus the relative contents of (1) 7–1 and (2) MDF-1 (arrows indicate the points of phase separation in the corresponding rubber–oligomer blends);

(*b*) versus the content of 7–1 in the oligomer compound at a fixed content of MDF-1 (percentage indicated at the curves);

(*c*) versus the total content of the oligomer compound at various component ratios MDF-1 : 7-1 (indicated at the curves).

were developed with a view to solving a different problem, namely, to reduce the viscosity of polyfunctional oligomers [22].

Predictions based on the structural-thermodynamic model were confirmed by the experimental data reported in [20]. As is seen from Fig. 4.3, using the approach outlined above we may calculate optimum compositions of the oligomer compounds and their contents in the blend, which ensure an increase by 1.5–3 times in the breaking strength of composites as compared to that of vulcanizates modified by individual oligomers.

4.3. CONTROL OF MECHANICAL PROPERTIES

The resistance to deformation (i.e., the elastic properties) exhibited by a system in response to the action of a force field is determined by the binding between structural elements of the system [14, 33, 102]. In cured oligomer blends, the degree of binding (i.e., the number of bonds per init volume) (i) is differently manifested on various levels of the structural hierarchy, (ii) depends on the total and relative contents of physical and chemical bonds between like and unlike components, and (iii) may differ significantly in various parts of macroscopic volume of the system. This pattern has two practically important implementations.

(I) The ability of atoms, atomic groups, segments, molecules, and supermolecular formations, entering into composition of a cured oligomer blend, to perform motions of various types in the course of deformation depends on the nature and structure of the continuous phase. The possibility to reveal the manifestations of various types of relaxation in blend composites depends (in contrast to the case of linear and cross-linked homopolymers) not only on the molecular and topological characteristics of structural elements and the regime of loading, but on the phase organization of the system as well, primarily on the matrix structure. This is a natural consequence, since it is the matrix in which the mechanical or other energy supplied from outside is transferred and distributed in the system. Let us consider two limiting cases. In a cured blend system, a high concentration of chemical network nodes is localized in the dispersed phase, while only a low concentration is found in the dispersion medium. This corresponds to a situation of glassy particles dispersed in a highly elastic (rubberlike) matrix. An inverse situation can be created in the cured system, whereby a continuous phase is highly cross-linked (glass) and contains dispersed particles with low-bonded chemical and physical nodes (vulcanized rubber). It is important to note that the deformation behavior of polymers occurring in the rubberlike and glassy states is described by different laws.[*] Although the system as a whole operates in the regime of deformation of the continuous phase, the elastic characteristics depend not only on the bulk ratio of components, but on the ratio of their moduli and the degree of interphase interaction as well.

Numerous variants of equations of the type [9] $P = aP_A^{n1}\varphi_A + bP_B^{n2}\varphi_B$ were proposed for the polymer blends and block-copolymers, but these are either not applicable to oligomer blends or valid only within a narrow range of compositions and temperatures.[**]

---

[*] These laws are considered in detail in [8–17, 33].

[**] In this equation P is a parameter (e.g., elastic modulus) characterizing some property of the blend system; P_A and P_B are the corresponding values

Incomplete phase and component separation (typical of the cured oligomer blends), different contents of the second component at a periphery (in the "transition" layer) of the nanoparticles of various types, nonuniform distribution of the nanoparticles over the material volume (the presence of micron particles representing agglomerates of the nanoparticles), and inhomogeneity of the continuous phase with respect to components are factors that hinder any quantitative generalization of the temperature and concentration dependences of the elastic deformation properties of these complicated systems at present. A few attempts of such quantitative assessment reported in the literature were not based on clear physical models and appeared as rather incorrect.

Nevertheless, we may well follow qualitative correlations between the topology, morphology, and phase organization of cured oligomer blends.

The first point to be noted is that phase inversion not only determines the character of the "stress-strain" $(\sigma - \varepsilon)$ diagram, but also accounts for a change in the character of the concentration dependence of elastic properties before and after inversion. It is the phase inversion, and the related change in the contributions of brittle and rubberlike deformations, that may account for the discussion (see [35]) around the experimental results reported by Klempner $et.$ $al.$ [34]. A characteristic shape of the concentration dependence of the deformation–strength properties (two humps with opposite signs), observed for the simultaneous interpenetrating network (IPN) systems based on polyurethane and polyacrylate (Fig. 4.4), is quite natural. Indeed, various relaxation states of the matrix (the elastomer–glass transition necessarily follows the phase inversion upon increase in the content of polyacrylate component) correspond to different influence of the dispersed phase on the elastic properties of the macroscopic system. These properties obey different laws above and below T_g.

The character of variation of the T_g value as a function of the composition, degree of binding (in particular, number of chemical network nodes), degree of component segregation, etc., in cured composites is rather complicated. Study of this character is of interest for both the fundamental science and (even more so, as demonstrated above) for the technological practice.

---

for the initial components A and B; φ_A and φ_B are the volume fractions of the components, respectively; a, b, n_1, and n_2 are coefficients determined by the blend structure, which may vary depending on the selected property and regime of testing.

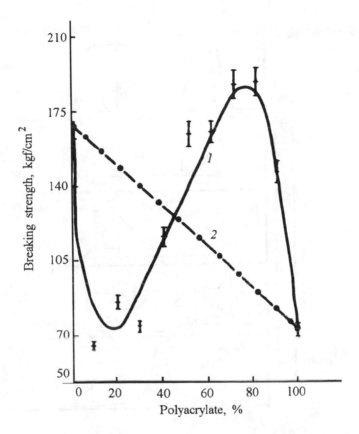

Figure 4.4. The plot of σ_b for polyurethane–polyacrylate IPNs versus the polyacrylate content [34]: (*1*) experiment; (*2*) calculation.

Methods used for the calculation and experimental determination of the T_g values of cross-linked polymers with various concentrations of chemical nodes were reviewed in [33]. The behavior is quite obvious: the α-transition shifts to lower temperatures with decreasing concentration of chemical bonds. In blend systems [8, 36] we can observe at least two T_g values corresponding to the α-transition for each of the components and several T_g values due to intermediate structures. The positions of these points depend on the extent of phase separation (or the degree of component segregation).

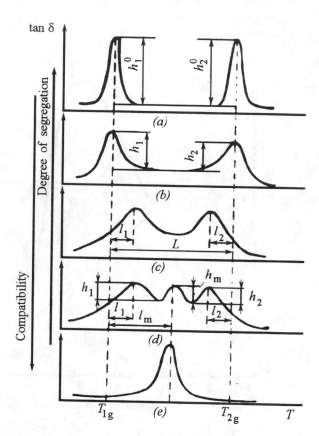

Figure 4.5. Temperature dependences of $\tan \delta$ in interpenetrating networks with various degrees of component segregation $\alpha = 1$ (a) 0.51 (b); 0.23 (c); 0.14 (d); 0 (e) [36].

Figure 4.5 shows an idealized behavior of $\tan \delta$ (loss tangent), used for the evaluation of T_g (by the position of $\tan \delta$ maximum), for IPNs with various degrees of component segregation [36]. Here, curve a corresponds to the complete phase separation (the degree of segregation $\alpha = 1$), whereby each phase is characterized by its own T_g value.

On the contrary, curve e in Fig. 4.5 corresponds to the case of components dispersed on the molecular level ($\alpha = 0$), forming solution with a single T_g value.

Intermediate cases (Fig. 4.5, curves b–d) reflect ralistic situations of incomplete phase separaton. The α value varies from 0.5 to 0.15; the glass transition temperatures of the components approach one another. As is seen, an additional T_g value, corresponding to the interphase region, appears at about $\alpha = 0.15$ besides the T_g values of the components (naturally, of the "deformed" ones).

The α values of systems with various extents of phase separation can be calculated using expressions derived in [36]. For the cases represented by curves b–d in Fig. 4.5, the corresponding formulas are as follows:

$$\alpha_b = \frac{h_1 + h_2 - \lambda_t}{h_1^0 - h_2^0},$$

$$\alpha_c = \frac{h_1 + h_2(\lambda_t - \lambda_m)}{h_1^0 - h_2^0}, \qquad (4.2)$$

$$\alpha_d = h_1 + h_2 - \left[\frac{h_1 l_1 + h_2 l_2 + h_m l_m}{h_1^0 - h_2^0} \right] \Big/ h_1 - h_2,$$

where $\lambda_t = (l_1 h_1 + l_2 h_2)/L$ and $\lambda_m = l_m h_m/L$ (the other notations as in Fig. 4.5).

(II) The second important consequence of variations in the degree of binding consists in the dependence of relaxation behavior of the macroscopic system on the ratio of fragments with various topologies in the structure of cured blends. The effects of various factors on the topology formation were considered in Sections 3.8 and 3.9. The role of the topological level of the blend structure organization in the mechanical properties was studied in much detail for the cured blends of oligomers with various functionalities and molecular masses [37].

Table 4.3 presents data [38] illustrating the effect of a monofunctional component on the properties of polyurethanes based on the blends of a tetrahydrofuran copolymer with propylene oxide, trimethylolpropane, 1,4-butanediol, and 2,4-toluylenediisocyanate. As is seen, an increase in the network defect density (growing with the fraction of monofunctional agent) is accompanied by a decrease in both modulus and strength levels. In practice, another important circumstance is that introducing a trifunctional agent into this mixture (with a view to increasing the average functionality reduced by adding the monofunctional molecules) by no means increases the properties, but may rather decrease the strength level (as confirmed by the data presented in [37]).

Promising results can be obtained by preliminary (before cure) drawing of an oligomer blend, which leads to the effect of molecular orientation of the reactive component (see Section 3.9), thus facilitating the formation of ordered structures in the cured oligomer

Table 4.3. Physicomechanical characteristics of cured polyurethanes with variable concentration of a monofunctional agent [38]*

$\rho_1 \cdot 10^2$	η_e/V, 10^{-4} mol/cm^3	σ_b, MPa	ε_b, %	E, MPa
0.053	3.67	2.25	140	1.80
0.085	2.82	2.10	160	1.50
0.130	1.29	1.05	210	0.75
0.147	1.12	0.85	190	0.60
0.158	1.02	0.70	190	0.40

Notes: *ρ_1 is the molar fraction of hydroxy groups in the monofunctional agent; η_e/V is the effective density of network.

Table 4.4. Physicomechanical characteristics of oriented polymer–oligomer blends [101]

System	Dose, kGy	σ_b, MPa	E, GPa	ε_b,%
PS + 50% DMEO-1	0/ 90	95/125	1.4/ 3.0	47.5/54.0
PS + 20% DMEO-3	0/ 38	270/125	6.1/ 2.6	49.0/43.0
PS + 50% DMEO-3	0/ 62	450/540	10.0/14.5	43.5/25.5
PS + 80% DMEO-3	0/ 51	45/ 90	2.0/ 1.2	79.0/52.0

blends. Table 4.4 gives physicomechanical properties of the blends of a high-molecular-mass polystyrene with n-oxyethylenedimethacrylates (DMEO) before (numerator) and after (denominator) the radiation cure [101]. The data were obtained for the optimum degree of drawing (see Table 3.3).

As is seen, the radiation cure of oriented systems improves physicomechanical properties of some compositions (for certain fluences [101]), while producing a negative effect for the other compositions (and doses). This non-unique influence of the orientation on the elastic-deformation properties indirectly confirms the activation mechanism of formation of a cybotaxic structure (see Section 2.8) and indicates that defectness of the network formed upon the cure may depend not only on the local order (anisotropy) of a reactive oligomer, but on the distribution of supermolecular formations in the sample volume as well.

Although the results described above have been given no satisfactory explanation within the framework of any commonly accepted (or discussed) model of the structure of cured oligomer blends, the last example has important practical implementations. It shows another unconventional method, namely, directed action of a mechanical field, for controlling (unfortunately, on an empirical level) the physicomechanical properties of composite materials based on oligomer blends.

An analysis of the entire body of data presented above, with an allowance for the results considered in Section 3.9, forms a basis for the development of methods to increase the stability of oligomer-blend-based materials with respect to alternating-sign stresses developed under conditions of cyclic (tension–compression) deformation.

First, note that the fracture of composite materials under the cyclic deformation conditions is a result of a combination of physical and chemical processes differing from those taking place in the course of uniaxial tension or compression. It is believed that the main difference between the processes of fracture under static and dynamic loading consists in the kinetic factors playing a dominating role in the latter case. Indeed, a certain part of structural elements does not attain an equilibrium state because the duration of a "loading–unloading" cycle is insufficient for realization of all the possible conformations of macromolecules and their segments corresponding to a given σ value.* At the same time, relaxation of the realized set of conformational rearrangements is also not completed by reaching a new equilibrium state. This leads to asynchronization of the process, formation of local overstressed microparcels, heat evolution, activated chemical degradation of the structure, and eventually, to accelerated fracture of the material.

In the first approximation, it was assumed [20] that the stability of heterogeneous oligomer-blend-based materials with respect to cyclic loading can be improved by increasing the variety of reversible and covalent bonds at the sites of structure primarily subjected to deformation, that is, at the periphery of globules and in the interglobular space. This would give rise to a system of "stress-absorbers", ensure an increase in the hysteresis losses, facilitate the energy dissipation, and reduce the probability of rupture of the covalent bonds. This structure can be obtained by partly loosening the newtork at the interphase boundary.

Experiments [20] confirmed that the introduction of agents (inhibitors) terminating the kinetic chain of polyreactions, responsible

---

* In this context we imply only reversible deformations. If the deformation energy is sufficient for breaking a certain part of chemical bonds in stressed (short or rigid) network fragments already during the first cycle, the mechanism of the fracture becomes much more involved.

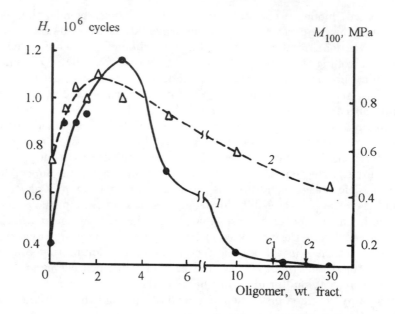

Figure 4.6. The plots of (*1*) fatigue durability H and (*2*) elastic modulus for the 100% deformation M_{100} versus the oligomer content in a cured blend of SKD–α,ω-methacrylate-bis(oxyethylene) phthalate. Arrows indicate concentrations corresponding to a phase separaton in the rubber–oligomer blend (c_1) and a maximum of σ_b (c_2); H is the number of cycles to breakage in the alternate-sign bending test.

for the curing of oligomer blends, leads to an increase in the dynamic durability H of the composites (as manifested by an extremum with respect to the inhibitor concentration), but at the expense of simultaneous decrease in the limiting breaking characteristics $[\sigma_b]$.

It was also established (Fig. 4.6) that the cure of a system comprising an elastomer matrix (capable of large reversible deformations) with reactive oligomers, forming dispersed high-modulus particles with additional bonds at the phase boundary, leads to a considerable increase in the dynamic durability of the material. However, the H versus c curve exhibits a maximum not coinciding with that on the σ_b vs c curve, which is quite natural in view of the different mechanisms responsible for the fracture of materials exposed to deformations of dissimilar types.

It might seem that, taking into account all the above considerations, there are no solutions that would allow us to simultaneously improve both the breaking strength and the dynamic properties of the oligomer blend composites. Nevertheless, proceeding from the scheme of formation of the structure of cured oligomer blends (see Section 3.9), it is possible to find some non-obvious technological methods capable of increasing the resistance to cyclic deformations, while retaining high limiting strength properties of the materials. Below we present several examples of using such methods.

Example A. A polymer–oligomer blend with the oligomer content equal to c_{crit} (a value ensuring the maximum σ_b, see Section 4.2) is modified by adding small amounts of a nonreactive plasticizer. The cure of this system involves microsyneresis of the inert molecules, which decreases the density of cross-linking at the periphery of primary globules. According to the above assumptions, concerning the structures with increased stability with respect to cyclic deformation, this material would exhibit a high dynamic strength. Indeed, introducing 3–5% dioctylphthalate into a system of SKD–30 wt.% MGF-1 leads to a vulcanizate having a high σ_b value, virtually equal to that of the nonplasticized blend (about 7.0 MPa), and the H value increasing about three times as compared to that for the latter system (from 3.1 to $9.5 \cdot 10^5$ cycles).

It is important to note that, using this method, one must select a nonreactive plasticizer well soluble in both rubber and oligomer components. No data are available for selecting the concentration of the additive.

Example B. A similar result (obtaining oligomer blends with increased durability under dynamic conditions) can be achieved by introducing reactive surfactant additives into the system. The efficiency of this approach is confirmed by experimental data presented in Table 4.2. Information concerning the effect of reactive surfactant additives on the rubber–oligomer blends was reviewed in [22], and on the oligomer–oligomer blends, in [26]. Note also that, in order to obtain a polymer–oligomer blend having optimum properties with respect to both σ_b and H, the oligomer content must be increased to $c > c_{crit}$. Indeed, the surfactant reduces the dimensions of segregated parcels of the phase of solution II, which would decrease σ_b if $c \leq c_{crit}$.

Example C. Linear polymers are modified by the compounds based on tetra- and polyfunctional oligomers. Here, the structure of cured blends usually (with a few exceptions [22]) has a more loose network as compared to that formed upon the cure of individual oligomers. This may lead to an increase in the dynamic durability:

systems presented in Fig. 4.3 not only possess high σ_b values, but exhibit a high H level as well [20].

4.4. CONTROL OF THE SHEAR VISCOSITY AND OTHER TECHNOLOGICAL PROPERTIES

Viscosity and other rheological parameters describe the elastic-deformation properties of materials. We will consider these characteristics in a separate section only because of their extremely high importance for the processing of reactive systems.

A process engineer concerned with development of an optimum process for the formation of thermoreactive systems has to simultaneously solve hardly compatible tasks: (i) obtain a composition possessing low viscosity at the stage of component mixing, (ii) maintain the low viscosity level in the course of processing up to the mold filling, and (iii) ensure a drastic increase in the blend viscosity upon the filling.

Important features of the rheological behavior of the initial oligomer blends were considered in Section 2.8. Variation of the viscoelastic characteristics of the blends in the course of cure can be divided into two stages. In the first stage, the blend viscosity increases as the polyreactions proceed. However, the system retains flowability because no common spatial network of covalent bonds, extended over the entire material volume, has been yet developed. In this stage, the process of curing (albeit initiated in a single-phase system) becomes microheterogeneous when the degree of polymerization reaches a certain level.* Here, the blend viscosity (initially described by the laws of solution flow) begins to obey the laws typical of flowing dispersions.

The second stage of the cure process begins on attaining a certain critical level of conversion β^*, which corresponds to the onset of gel formation and the loss of flowability. The curing is completed and the final structure is formed in a macrogel phase (see Section 3.8).

Figure 4.7 shows a generalized diagram representing the kinetics of variation of the fraction of unreacted groups N, the viscosity η, and the elastic modulus G_∞ in the course of three-dimensional condensation according to Flory.

The classical theory of gel formation [40], which considers the case of formation of an ideal network in the course of polycondensation

---

* The particular values of conversion, at which the system exhibits the "macrophase" transitions, depend on the nature of components, the mechanism of polyreactions, and the regime of cure kinetics (see Section 3.7).

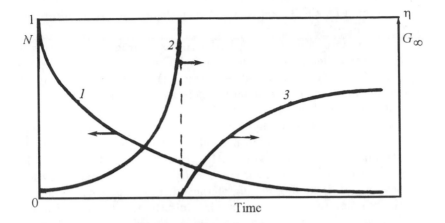

Figure 4.7. Kinetics of (*1*) the loss of functional groups *N* and the increase in (*2*) viscosity η and (*3*) elastic modulus *G* in the course of the three-dimensional condensation according to Flory [40].

(limitations of the initial model employed), leads to simple relations between the coefficient of branching α^* at a given point of the gel and the β^* value for the reaction $A + B \rightarrow A - B$:

$$\alpha^* = \frac{1}{f-1}, \quad \alpha = \frac{r\beta_A^2 \rho}{1 - r\beta_A^2(1-\rho)} = \frac{\beta_B^2 \rho}{r - \beta_B^2(1-\rho)}, \qquad (4.3)$$

where f is the arithmetic mean value of the functionalities of molecules A and B, r is the stoichiometric coefficient (the $A : B$ molar ratio), and ρ is the fraction of reactive groups of a given kind, entering into polyfunctional molecules, among the total number of these groups in the system.

For example, using Eq. (4.3) we can readily calculate that interaction between two trifunctional compounds ($f = 3$, $\alpha^* = 0.5$) taken in the equimolar ratio ($r = 1$, $\rho = 1$) corresponds to $\beta_A^* = \beta_B^* = \beta^* = 0.707$.

However, numerous experimental β^* values reported by various authors have never coincided with the calculation. Repeated attempts to reach agreement between experiment and theory by introducing additional empirical coefficients and new parameters (taking into account cyclization and the possibility of forming other topological structures) in the Flory equations were insuccessful. Factors that account for the discrepancy between theory and experiment were analyzed in detail, e.g., in [41].

The pattern of variation of the viscoelastic properties depends on the particular details of topology and the conformation of branched (sol fraction) and cross-linked (gel fraction) structures. Special features of the rheokinetics of particular oligomer systems were thoroughly considered in [42].

The rheokinetics of oligomer blends was studied predominantly for the process of urethane formation involved in the injection molding technology (PIM process). At the present stage of knowledge, this complicated set of problems is empirically solved for separate technological stages, including mixing of the initial components, transport of the reaction mixture to the mold, filling the mold, cure of the material, and cooling of the article. The results of investgation in this direction are summerized in [43–45, 103].

4.5. INFLUENCE OF ADDITIVES ON THE PROPERTIES OF OLIGOMER-BASED MATERIALS

4.5.1. Fillers

The problem of polymer composite filling implies a variety of complicated tasks. There are several well-known monographs covering only the physicochemical aspects of this problem [6, 14, 46, 47]. However, many important questions still remain unclear.

The circle of unresolved problems is even more extended in the case of oligomer blends because of a complicated organization of these systems. Technologists and researchers engaged in oligomer blends have to meet, in addition to the traditional problems, a number of new ones specific for the systems. These include, in particular, selectivity of the interaction of a filler with the blend components, influence of the filler on the compatibility of components, distribution of the filler in coexisting phases, influence of the solid surface on the cure kinetics, role of filler particles as nuclei for the phase separation in the course of the cure, and many others.

Below we formulate and briefly consider only some of these problems in application to the oligomer blends with highly disperse fillers.*

---

* It should be noted that problems related to the fiber-reinforced oligomer blend composites are very special and require separate consideration.

Displacement of the concentration equilibrium. Introduction of a filler N into a thermodynamically equilibrium mixture of two components, A and B, leads to displacement of the equilibrium and redistribution of bonds in the oligomer blend. In the first approximation, an analysis of the thermodynamics of these systems reduces to the following.

The free energy of mixing A and B is given by the relation $\Delta G_{AB} = RT\chi_{AB}\varphi_A\varphi_B$, where χ_{AB} is the thermodynamic interaction parameter of the components, φ_A and φ_B are their volume fractions, R is the gas constant, and T is the absolute temperature.

The free energies of mixing of each component with a filler can be written in a similar form: $\Delta G_{AN} = RT\chi_{AN}\varphi_A\varphi_N$ and $\Delta G_{BN} = RT\chi_{BN}\varphi_B\varphi_N$, where χ_{AN} and χ_{BN} are the interaction parameters of each component with the filler and φ_N is the volume fraction of the filler.

On introducing a filler into a system, we can distinguish between two variants: $\Delta G_{AN} \approx \Delta G_{BN}$ and $\Delta G_{AN} \neq \Delta G_{BN}$. In the former case, the equilibrium relative concentration of blend components near the filler particles and in the bulk remains unchanged, while in the latter case the situation is altered.

In the second case, for $\Delta G_{AN} < \Delta G_{AB}$ (or $\Delta G_{BN} > \Delta G_{AB}$) the region in the vicinity of a filler particle becomes enriched with component A (and the bulk is accordingly depleted of this component). For $\Delta G_{AN} > \Delta G_{AB}$ (or $\Delta G_{BN} < \Delta G_{AB}$), the situation is reversed. However, the condition $\Delta G_{AN} \neq \Delta G_{BN}$ always implies that introduction of the filler leads to redistribution of the components. As a result, concentration of one of the components in the bulk decreases, while that in the boundary layer increases.

Experimental data showing evidence of the redistribution of components caused by the introduction of fillers into oligomer blends were reported in [46–51]. Note, in particular, that adding a filler to a blend of epoxy oligomer and polybutadiene rubber [51] led not only to a local change in the oligomer concentration, but to a modification of the conformation of macromolecules in the boundary layer as well.

According to the physical meaning (and definition) of the parameter χ_{AB}, this quantity must not depend on the presence of a filler. However, some data [5, 8, 20, 70] unambiguously indicated that χ_{AB} varied with the amount of filler and the A to B concentration ratio. There are various possible explanations for this fact, of which the following seems to be the most reasonable. Oligomer blend systems are predominantly the mixtures of homologous oligomers and polymers, differing by their molecular masses. It was demonstrated that adsorption on the solid surface of the filler leads to concentration of fractions with lower molecular mass. In other words, the filler absorbs primarily and mostly the low-molecular fractions of components A and B. Accordingly, the proportion of high-molecular fractions of both components (or at least one of them in the case of selective adsorption) inreases in the bulk (i.e., at a sufficiently large distance from the filler surface). This circumstance inavoidably leads to a change in the compatibility of components A and B, as reflected by variation of the χ_{AB} value and the interaction parameter depending on the filler content and the initial mixture composition.

The change in the compatibility of components in the bulk and the concentration of one component in the boundary regions at the filler surface obviously modify the kinetics and structure, which can be readily followed by analysis of the scheme of processes involved in the cure of oligomer blends (Sections 3.8 and 3.9).

Effect of the spatial distribution of filler particles. Uniform distribution of the particles of filler in the volume ensures the best properties of a filled material [11–14]. One of the reasons was considered above in Section 4.2, and another reason consists in that regions enriched with the filler may serve as a source of brittle fracture. Principles of the quantitative description of a relationship between the spatial distribution of filler particles and the properties of materials are not yet completely formulated. Tovmasyan [52] theoretically analyzed the distribution of particles in agglomerates and experimentally verified the model for the dispersion of glass spheres in polyethylene with a low (15%) filler content. No such investigations have been attempted for the oligomer blend systems. However, it is evident that the agglomeration of filler particles, leading to the formation of coarse randomly distributed inclusions, may decrease (or even eliminate) the reinforcement effect of the filler.

The problem of "homogenization" of the filler distribution in polymer systems, albeit an apparently simple formulation, is essentially contradictory. Technologically, the task is usually solved at the expense of higher energy consumed for the mixing of blend components. This is achieved by increasing the rotation speed of elements of the blending equipment, reducing the clearance between rollers, and increasing their friction (e.g., by increase in the difference of rotation speeds), which usually leads to negative effects such as mechanodestruction. Introduction of special additives, to serve as dispersing agents, is also not always expedient.

In some cases, the problem of uniform distribution of fillers in polymer–oligomer blends can be solved by unconventional methods [20]. The unusual approach consists in increasing the viscosity of the blend at a certain stage of mixing, instead of a decrease in the polymer viscosity provided in the conventional processing of filled systems.* The effect is achieved using a simple technological method, namely, by changing the order of component introduction.

The idea of the method is essentially as follows. The degree of agglomeration of the filler and the character of its distribution in the system depend in a rather complicated manner (not yet described by hydrodynamic methods) on the viscosity of matrix (as described in more detail in [53]). It was empirically established that the optimum distribution of a filler, ensuring the maximum σ_b value of the vulcanizate, can be achieved only within a rather narrow interval of values of the rubber plasticity, where the agglomerates of filler particles are destroyed on mixing with the polymer component and the probability of clotting is excluded. Using rather involved dependences of the viscosity of rubber–oligomer blends on the nature and content of the rubber, we can control the rubber plasticity in the course of the process.

For example, let us consider the introduction of a filler (technical carbon) into a cis-polyisoprene rubber modified by oligoester acrylates. The blending is performed in a roll mill or in a rubber mixer. In the first step, a linear oligomer (0.5–1.5% of TGM-3, MGF-9, MDF-1, etc.) is added to the rubber, which is accompanied by a decrease in the plasticity (see Section 2.9). In the second stage, the carbon component (soot) is introduced and dispersed in the high-viscosity matrix. It is important that the matrix (i.e., the rubber–oligomer blend) initially (by the moment of filler introduction) possesses an elevated viscosity, which grows further on introducing the soot. Then, the carbon-filled mixture is modified by adding either an excess of the same linear oligoester acrylate,** or a branched oligomer (of the 7-1 or 7-20 type). An increase in the total content of oligoester acrylate leads to a sharp drop in the viscosity of the mixture, in accordance with the classical laws of polymer plasticization. This facilitates mixing of the rubber blend with other components and compensates the excess energy consumed to mix the high-viscosity mass in the first stage of blending. It should be emphasized that reduction in the viscosity in the second blending stage does not affect the structure of filler agglomerates formed in the first stage.

---

* The viscosity can be reduced for various purposes, in particular, for decreasing the energy consumed for the blend processing.

** The amount of oligomer in the second portion is determined from the $\eta = f(c)$ or $\sigma_b = f(c)$ dependences.

Table 4.5. Effect of the regime of oligoester acrylate introduction on the properties of carbon-filled peroxide vulcanizates based on the SKI-3 polyisoprene rubber.* Regime I: rubber + TGM-3 + filler + 7-1; regime II: rubber + (TGM-3 + 7-1) + filler

Property	Oligomer content**, mol.%												
	0.5	0.5	1.5	2.0	5.0	1.0	0.5	0.5	1.5	2.0	5.0	1.0	0
	0.5	1.0	0.5	2.0	1.0	5.0	0.5	1.0	0.5	2.0	1.0	5.0	0
	Regime I						Regime II						
σ_b,MPa	20	22	19	15	15	17	18	19	18	15	17	12	16-17
M_{300}, MPa	11	13	10	8	8	8	9	10	11	8	7	4	8-10
l,%	550	500	500	500	450	400	550	550	550	500	450	400	500
Energy consumption,*** rel. units	0.9	0.7	0.8	0.7	0.7	0.7	0.9	0.7	0.8	0.6	0.6	0.6	1.0

Notes: *Filler, 30 mol. fract. of the PM-50 commercial carbon (soot) per 100 mol. fract. of the rubber.
**Oligomer content: numerator, TGM-3, denominator, 7-1.
***Relative to the energy consumed for the mixing of rubber with soot in the absence of oligomer.

The above procedure ensures the obtaining of vulcanizates with increased strength level at the same (or even reduced) total energy consumed for the elastomer blend preparation.

Data presented in Table 4.5 show that alteration in the order of oligomer and filler introduction, together with a change in the relative oligomer content, may completely or partly compensate for the effect of vulcanizate strengthening achieved at the expense of better dispersion of the filler.

Effect of the degree of filler dispersion. An important role of this factor was partly considered above. First, the limitation of agglomeration is one of the methods to control the dispersion of filler and, hence, the system morphology. Second, variation of the dispersity and content of the filler allows us to control the phase organization of the system. Indeed, the dispersity determines the specific surface area of the filler and, hence, the adsorption capacity of the filler and the equilibrium concentrations in the system.

The role of the degree of filler dispersion in an oligomer blend can be also analyzed from a different standpoint, taking into account the heterogeneous nature of cured blends. The microheterogeneous particles formed in the course of cure can be considered as particles or agglomerates of an active polymer filler embedded into a matrix of another polymer, namely, of the initial polymer (in the case of polymer–oligomer blends) or the polymer phase formed upon the cure (oligomer–oligomer blends). The use of additional mineral fillers in these systems markedly extends our ability to control the topology and phase structure of composites. Note that both small ($< 10\%$) and very large ($> 80\%$) contents of a filler can be used to reach certain effects. However, obtaining a desired structure and a required combination of properties may require correcting the initial oligomer blend composition and observing definite requirements with respect to the size of mineral filler particles.

For example, obtaining the maximum σ_b value may require that the upper size of the filler particles would not exceed $D_{\lim}$ (as in the case of unfilled oligomer blend systems, see Section 4.2). While the unfilled composites with maximum strength are obtained only upon the cure of two-phase initial blends, the introduction of a filler may shift the optimum ratio of components in the matrix toward a single-phase state. In other words, the filler may lead to lower oligomer consumption, while retaining a high level of the strength properties of cured composites.

As a rule, the size of mineral filler particles used in the practice is of the same order of magnitude as that of the micron particles formed upon the cure of two-phase oligomer blends. Therefore, it is not necessary to cure two-phase blends in order to fulfill the requirement of a wide particle size distribution in the case of filled systems, because it is the right-hand "tail" of the distribution function ($0.1\ \mu\text{m} \leq R < 10\ \mu\text{m}$) that ensures the introduction of mineral filler particles into the oligomer blend system. At the same time, with this approach to the creation of composites with preset properties it is evident that introduction of reactive oligomers into filled polymer blends ensures the formation of nanoparticles upon curing. This implies that we "fill" the part of the size distribution function ($R < 0.1\ \mu\text{m}$) that is always "empty" in the usual filled polymer systems, despite polydispersity of the conventional mineral fillers. This is one of the factors explaining the efficiency of using the reactive additives for the modification of filled polymer–oligomer systems.

A process engineer selecting the concentrations of components in filled oligomer blends must also take into account another important factor mentioned above. A certain proportion of the reactive oligomer is sorbed by the filler surface and is therefore excluded from the bulk polymerization process. Thus, the bulk is depleted of the nanoparticles formed upon curing, which may decrease the strength level of the composite. In order to compensate these losses, it is necessary to correct the initial composition for the adsorption capacity of the filler, which is determined by its content, dispersity, specific surface area, "affinity" to the oligomer, etc.

Other special features. The adsorption of reactive oligomers on the filler surface has several positive aspects.

According to the well-known classification of polymer–filler interactions [14], the filled systems are subdivided into four groups: (i) a simple mechanical mixture of polymer and filler, which merely leads to dilution of the system and decreases the strength of composite; (ii) wetting of the filler surface by the polymer increases the interaction and, accordingly, improves the properties of the composite; (iii) physical contacts between phases, which produces a still greater effect; and (iv) chemical bonds at the phase boundary — the ideal to be approached.

Although the conclusions made in terms of this scheme are too categorical, the adhesion obviously plays an important role in the mechanical properties of composites. This factor is extremely important for the strengthening of filled composites with respect to any types of deformation and operative fracture mechanisms, as is well illustrated by the well-known results of Sato and Furukawa [54] obtained originally for the filled elastomers (Fig.4.8) and later confirmed for the filled matrices of other types [14, 55].

As is seen from Fig.4.8, the strength of a material increases with the filler content in the case of the "ideal" adhesion of polymer to filler ($\xi = 1$) and decreases in the absence of adhesion bonding ($\xi = 0$). Intermediate values of the adhesion parameter ($0 < \xi < 1$) determine the level of strength properties in the corresponding interval of filling.

Within the framework of the above classification, the filled oligomer-blend materials can be assigned to the third and fourth groups. Indeed, the reactive groups occurring on the filler surface may interact with adsorbed oligomer particles in the presence of appropriate functional groups and conditions (initiator, temperature, etc.). However, even if no chemical interactions are possible in the near-surface regions, the adsorption on the filler surface increases the density of physical bonds and improves adhesion at the filler–matrix interface.

Thus, the cure of filled oligomer blends (even without any special pretreatment of the filler) may allow us to perform, in principle, two

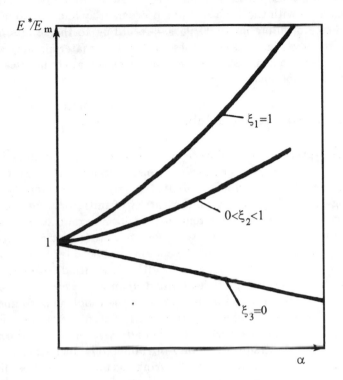

Figure 4.8. Concentration dependence of the elastomer strengthening coefficient E^*/E_m calculated for various values of the adhesion parameter ξ [54].

most commonly employed methods of filler modification — dressing and capsulation. In practice, this reduces to controlled alteration in the order of component mixing, variation of the temperature regime of initiation, and strict observation of the cure time schedule.

Another promising possibility of the oligomer blend filling that is also worth mentioning is the use of finely powdered secondary polymeric materials as fillers. Introduction of these fillers at a large amount into melts and solutions of primary polymers may be difficult, inexpedient, or even undesirable because the resulting blend systems would belong only to the first (or, in the best case, to the second) group of materials. However, the situation may essentially change on introducing these fillers into oligomer blends. The active interaction of oligomers with the surface of fine powders may allow us to increase the degree of filling up to 90–95%. According to the results of preliminarty investigations [56], these superfilled materials may exceed with respect to some parameters the conventional composites based on primary polymers.

4.5.2. Activators and Inhibitors

Some aspects of the influence of activators and inhibitors on the structure of cured composites were considered above.* In Section 4.1 we have analyzed possible trends in the variation of composite morphology depending on the concentration of initiators and inhibitors in oligomer blends with different initial phase organizations. Section 3.7 presents the results [70] clearly demonstrating that an increase in concentration of a catalyst for the epoxide–rubber blend cure reaction leads to a change in the shape and width of the size distribution function of heterogeneous particles formed upon cure. In Section 4.3 we have mentioned the effect of inhibitors on the topology of composites, whereby introduction of a chain terminating agent leads to the formation of a less dense network in the boundary layer, thus ensuring an increase in the dynamic durability of rubber–oligomer vulcanizates.

The effects of catalysts and curing agents on the cross-linking density and some topology characteristics of composites, obtained in urethane and epoxide systems, and the resulting influence of these additives on the elastic-deformation properties of these systems in the glassy and rubberlike states were described and correctly analyzed in [33, 37, 59]. The results of systematic investigations of the inhibited polymerization of unsaturated oligomers were presented in [58, 63]. However, the volume of data related to the effects of activators and inhibitors on the cure process in oligomer blends is still insufficient.

---

* By activators we imply all the chemical reagents and energy factors capable of reducing the energy barriers for chemical reactions, thus "switching on" the mechanisms of polyreactions leading to the cure of blends. The class of activators includes chemical compounds (initiators of radical and ionic processes, curing agents, and catalysts of the polycondensation and migration polymerization processes), thermal and light energy, irradiations, etc. Obviously, these factors have different mechanisms of action. The mechanisms are described in monographs and reviews devoted to particular oligomer systems [33, 57–60].

Below we will consider only some important examples of the combined action of activators and inhibitors on the process of oligomer blend curing.

Anaerobic Systems. The ability of (meth)acrylic bonds to be stable in air, changed by their rapid polymerization upon isolation from the atmosphere, predetermined the use of oligoester acrylates in anaerobic sealer compositions [71].

Anaerobic sealer compounds are multicomponent systems containing several necessary components (acrylic oligomers, monomers, activators, and inhibitors) and some additives (modifiers, plasticizers, thickeners, complex-formers, etc.) controlling special properties of the compositions.

In the presence of oxygen (e.g., in air), the activity of inhibitors prevails over that of initiators and the system occurs in the liquid state. In order to maintain anaerobic sealer compounds in the working condition before use, they are stored in polyethylene containers whose walls are readily permeable for oxygen molecules. A critical thickness of the layer of sealing compound, below which the reaction switches to the anaerobic regime, depends on a number of factors, primarily on the efficiency and concentration of inhibitor.

In some time after introducing an anaerobic sealer into a closed volume with no air access (e.g., into a thread clearance between screw and nut, a crevice in metal, etc.), namely, on expiry of an induction period required to consume the residual oxygen, the system passes into the anaerobic regime and the cure process begins.

A variant of the theory of anaerobic polymerization was developed by Tvorogov *et. al.* [72, 73] on the basis of theories of the classical inhibited polymerization [74] and inhibited oxidative polymerization [63].

The main relationships determining the induction period duration, the degree of polymerization, and the critical layer thickness l_{cr} (at which the liquid system can be stored) are as follows:

$$t = \mu[\text{In}]_0/v_i$$

$$(c_0 - c_t)/c_0 = 1 - ([\text{In}]_t/[\text{In}]_0)^{k_p/k_{in}} \qquad (4.4)$$

$$l_{cr} = 1.4\left[\frac{D[\text{O}_2]_0}{v_{\text{O}_2}}\left(1 - \frac{[\text{O}_2]_{cr}}{[\text{O}_2]_0}\right)\right]^{1/2} \approx 1.4(D \cdot t_{cl})^{1/2},$$

where v_i is the initiation rate, c_0, c_t, $[\text{In}]_0$, and $[\text{In}]_t$ are the initial and current concentrations of the reactive component and the inhibitor, respectively, μ is the stoichiometric coefficient, k_p and k_{in} are the rate constants of the chain propagation and termination (inhibilion) reactions, respectively ($R^{\bullet} + M \rightarrow R^{\bullet}$, $R^{\bullet} + In \rightarrow In^{\bullet}$ or $RO_2^{\bullet} + M \rightarrow R^{\bullet}$, $RO_2^{\bullet} + In \rightarrow In^{\bullet}$, depending on whether the reaction proceeds under anaerobic or aerobic conditions), v_{O_2} is the rate of oxygen consumption in the reaction of oxidative polymerization, $[O_2]_0$ and $[O_2]_{cr}$ are the initial and critical concentrations of oxygen, at which the polymerization regime switches from inhibited to uninhibited with respect to oxygen, D is the diffusion coefficient, and t_{cl} is the induction period in the closed system.

Equations (4.4) show that increasing inhibitor concentration and decreasing initiation rate lead to longer induction period in the presence of oxygen (the maximum working life t_w). However, the induction period in the closed system t_{cl} (i.e., under the working sealer conditions) and the degree of polymerization vary in the opposite directions. In other words, a longer working life of the open (aerobic) system corresponds to lower working properties of the sealer (a greater time of the cure under anaerobic conditions) and lower final properties of the cured composite.

Two approaches were suggested to provide for a combination of sufficiently long working life of the system and comparatively rapid curing. The first method consists in introducing large concentrations of low-efficient (weak) inhibitors, and the second, in using small concentrations of strong inhibitors together with additives restoring the oxidized (working) form of the antioxidant.

Table 4.6 presents the results of a systematic investigation [73], showing possibilities of both approaches. The main conclusions from this table are as follows. The ratio of the parameters corresponding to satisfactory storage and working conditions is independent of the nature of initiator.

For comparable values of the induction period under the storage conditions t_{st} (open system), the induction periods in the curing regimes (closed system) t_{cl} are markedly lower for weak inhibitors, such as orthohydroxyquinoline (OHQ) or dichlorohydroxyquinoline (DCHQ), than for a strong one (bisphenol 22-46).

It should be noted that, besides the positive effect, the use of weak inhibitors leads to undesired decrease of l_{cr} during storage.

Characteristics of the main commercial anaerobic sealer compounds are given in [71, 75]. Sineokov et. al. [76, 77] described new polyfunctional (meth)acrylic oligomer compositions and the corresponding activator–inhibitor systems, which allow the processing and working properties of these anaerobic sealers to be varied within wide limits.

Table 4.6. Process parameters and storage characteristics of anaerobic sealer systems in the presence of various inhibitors [$v_i = 9.42 \cdot 10^{-11}$ mol/(l·s); $T_{st} = 20°C$; $T_{cure} = 60°C$].

Inhibitor	[I]$_0$, mol/l	[In], 10^{-3} mol/l	t_{st}, years	t_{cure}, min	t_{wl}, years	l_{cl}, cm
OHQ	PB	–	0	4.0	–	2.0
	$(1.8 \cdot 10^{-2})$	3.4	0	4.0	1.15	3.3
		3.4	0.5	4.0	–	1.9
		11.0	0	8.0	3.7	4.9
		11.0	1.0	5.0	–	3.7
		11.0	2.0	3.0	–	3.0
		29.0	0	17.0	9.8	7.0
		29.0	2.0	14.0	–	6.3
DCHQ	AIBN	–	0	6.0	–	2.0
	$(1.8 \cdot 10^{-3})$	10.0	0	23.0	3.4	8.0
		10.0	1.0	16.0	–	7.0
		70.0	0	115.0	23.0	18.0
		70.0	1.0	110.0	–	18.0
		70.0	5.0	108.0	–	18.0
Bisphenol 22-46	AIBN $(4.1 \cdot 10^{-3})$	3.6	0	470.0	1.2	37.0
		3.6	0.5	430.0	–	35.2
		13.0	0	638.0	4.4	43.0
		13.0	1.0	576.0	–	41.0
		13.0	2.0	557.0	–	40.0

Inhibitors of the composite degradation. One of the most difficult tasks to be solved in the development of oligomer blend materials cured by the radical mechanism is to introduce inhibitors (stabilizers or antioxidants) capable of preventing degradation of the material during the subsequent thermal and thermooxidative aging [78].

Obviously, it is virtually impossible to introduce the stabilizing additives into the final structure of nonmelting and insoluble composites. At the same time, it is inexpedient to introduce the highly efficient inhibitors into the initial compositions, because this may suppress the chain propagation in the course of the radical polymerization and hinder the formation of a cross-linked structure.*

---

* There is no sense in surmounting the inhibiting activities of antiox-

This problem can be solved, in principle, by using antioxidant additives that are inefficient (or low-efficient) as inhibitors of the radical processes at temperatures used for the material molding, but are efficient with respect to the same processes at other temperatures (e.g., under the material exploitation conditions).

An example of realization of this approach is offered by the results of a screening test among more than 50 compounds, belonging to various classes of inhibitors of the amine and phenol types [79]. It was found that mixtures of alkylphenoxypropylene oxide, 1,3-dialkylesters of glycerol, and alkylphenols (PF-3) or the product of an epoxidiane oligomer etherification by tetramethylenehydroxypiperidine (PF-7), introduced at an amount of not exceeding 1%, do not virtually suppress the radiation curing of the epoxidiane oligomer at room temperature. Moreover, these additives are capable of activating the three-dimensional polymerization under certain conditions. At the same time, these additives very efficiently inhibited the thermooxidative destruction of cured composites. Figures 4.9 and 4.10 present experimental evidence to confirm this statement.

Study of the shear viscosity of compositions with and without these inhibitors, together with an analysis of the kinetics of initial oxidation rates of inhibited and uninhibited compositions, showed [79] that the addition of inhibitors (e.g., PF-3 and PF-7 selected by screening) facilitates the formation of structures with labile templates possessing "kinetically favorable" arrangement of reactive oligomers in liquid oligoepoxy acrylate compositions (see Section 3.6). Rapid polymerization in these structures compensates losses in the rate and degree of polymerization, which are caused by recombination and chain termination processes induced by the same inhibitors. As a result, the presence of PF-3 and PF-7 ensures the obtaining of solid cross-linked structures with satisfactory strength properties, which are much more stable with respect to thermooxidative destruction as compared to the composites obtained without these stabilizers.

Note that mechanisms outlined above, albeit not strictly confirmed, provide a quite reasonsble explanation of the observed effects.

Radical polymerization in matrices possessing high inhibiting activity. The most typical examples of such systems is

---

idants during the initiated radical polymerization of oligomer blends by analogy with the anaerobic sealer compounds. Indeed, the inhibitor is "deactivated" during the induction period and has no effect on degradation of the cross-linked polymer.

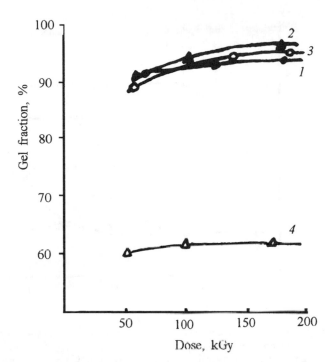

Figure 4.9. The content of gel fraction versus the absorbed radiation dose in cured epoxyacrylic oligomers based on (*1*) diglycidyl ester of oligoepichlorohydrin (ADEE) and its mixtures with the products of the epoxy oligomer etherification by (*2*) 2,2,6,6-tatremethyl-4-hydroxypiperidine (PF-7), (*3*) alkylphenoxypropylene oxide (PF-3), and (*4*) 1,3-di(*p*-phenylaminophenoxy)-2-propanol [79].

offered by bitumens modified by polymerizable oligomers. A difficuly in obtaining a positive result (i.e., a final material with increased properties) consists in that we can by no means always ensure polymcrization of the unsaturated compounds introduced into bitumens. Indeed, the asphaltene fractions of petroleum bitumens, composed by most part of aromatic compounds with a polyconjugated system [80], may (and really do) inhibit the radical processes involved in the cure of modifying agents. In order to surmount the action of asphaltene inhibitors, the bitumen–oligomer blends are sometimes additionally modified by adding large amounts of initiators. According to [81], the cured bitumen– oligoester acrylate blends with optimum properties

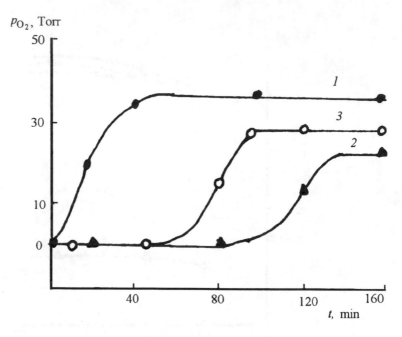

Figure 4.10. Kinetics of oxygen absorption for the oxidation of radiation-cured oligomers: (*1*) ADEE; (*2*) ADEE + PF-3; (*3*) ADEE + PF-7. Oxidation temperature, 170°C.

are obtained at a tertbutyl peroxide (peroximon F-40) concentration of 5–7%. According to some other data [82, 83], the properties of bitumen–oligomer compositions were improved at a much lower initiator contents (0.5–1%).

It was shown [84] that dependence of the relaxation and strength properties of the cured bitumen–oligomer systems on the curing temperature T_{cure} is described by curves with extrema such as depicted in Fig. 4.11. As is seen, blends composed of a BND-60/90 bitumen with oligoester acrylates and oligoepoxy acrylates, cured in the temperature interval $140°C \leq T_{cure} \leq 160°C$, exhibit a significant increase in the softening temperature T_{soft}, which is caused by intensive chemical conversions leading to the formation of cross-linked structures. As is seen from Fig. 4.11, the content of oligomer and the nature of initiator determine only the peak magnitude and the position of its maximum on the temperature scale, while the shape of the extremal curve $T_{soft} = f(T_{cure})$ remains unchanged.

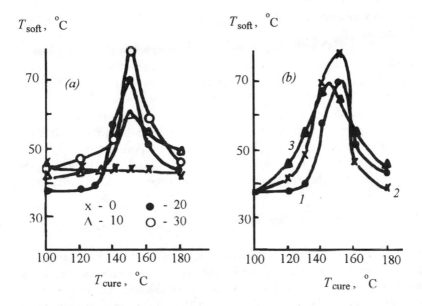

Figure 4.11. The plots of T_{soft} versus T_{cure} (curing time, 2 h).
(a) For ($\times$) a BND-60/90 bitumen and its blends with ($\triangle$) 10%, ($\bullet$) 20%, and (o) 30% trioxyethylenodimethacryl cured in the presence of dicumyl peroxide;
(b) For a bitumen–20% oligomer blend cured in the presence of various initiators: (*1*) dicumyl peroxide; (*2*) benzoyl peroxide; (*3*) azobutyronitrile of isobutyric acid.

This behavior is related to the possible manifestation of the effect of "local activation" inherent in compounds with a polyconjugated system of bonds [85, 86]. This effect consists, in particular, in that the polyconjugated compounds may act (in some cases, determined by the conjugation length, localization of the electron density, potential barrier, temperature, etc.) as sources of active radicals (or the promotors of radical formation), while in other cases (with different parameters of state) these compounds may be neutral with respect to radicals or even inhibit their activity.

It was suggested [84] that aromatic polyconjugated compounds, contained in the asphaltene fraction of bitumen, exhibit inhibiting properties at $T_{cure} > 160°C$. There are various possible mechanisms of this effect. For example, under these conditions polyconjugated compounds may serve as the traps for radicals formed upon the peroxide decomposition. As a result, the chain radical processes would be suppressed.

At $T_{cure} \leq 130°C$, the polymerization processes are suppressed because the efficiency of initiation (i.e., the concentration of radicals formed upon the thermal decomposition of peroxide) is insufficient to "switch on" the process of chain polymerization process.

Thus, it is only within an intermediate temperature interval that the chain propagation rate exceeds the rate of radical recombination and termination reactions. Indeed, in this interval either the "poly-conjugation length" is insensitive to peroxide radicals or the structures formed are capable of activating the development of chain reactions.

4.5.3. Nonreactive Plasticizers

It is only in the stages preceding the cure that we can introduce low-molecular-mass plasticizers (softeners) into polymer materials in which high cross-linking densities are obtained upon the cure of initial oligomer blends. In order to plasticize the final cross-linked polymer (i.e., to impart elasticity to the structure), it is necessary to incorporate plasticizer into the composition. However, the plasticizer affects the thermodynamics of mixing of the main components, the supermolecular organization, and the phase structure of the mixture, thus changing its molding properties and cure kinetics, and, hence, the properties of final composites.

No systematic investigations have been undertaken by now that would concern all the above-mentioned aspects of the effect of nonreactive plasticizers on the oligomer blend structure evolution. However, some parts of this problem were studied in sufficient detail.

Study of structural transformations in epoxy and furan oligomers [87] showed that low-molecular-mass plasticizers may significantly affect the dimensions of inhomogeneities formed in the so-called associated liquids to which these oligomers belong. Depending on the nature and content of the plasticizing additive, it may either increase or decrease the size of supermolecular formations. It was found that the density, viscosity, plasticity, and other properties of uncured compositions vary in accordance with changes in the suermolecular structure.

Upon curing, the parameters of morphology (i.e., the dimensions and volume fraction of globules) vary as functions of the plasticizer content and obey the same laws as the size of associates in the uncured blends. It was also established that the low-molecular-mass additives tend to localize predominantly in the defect (low-cross-linked) regions

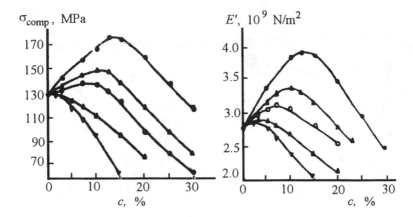

Figure 4.12. The plots of (*a*) σ_{comp} and (*b*) E' versus plasticizer concentration *c* for the cured composites based on the epoxy oligomer ED-20 plasticized by (*1*) pentachlorodiphenyl, (*2*) trichlorodiphenyl, (*3*) dichlorodiphenyl, (*4*) diphenyl, and (*5*) dibutylphthalate [87].

of the composite. If the plasticizer has a chemical structure ensuring high thermodynamic affinity (i.e., compatibility) with respect to the initial oligomer and, naturally, retains the affinity to structural elements of the cross-linked polymer, the resulting composite is characterized by higher packing density in the interglobular space, lower free volume, narrower conformation spectrum, etc.

In some cases, the above combination of factors leads to an increase in the equilibrium and deformation–strength properties of cured composites (this phenomenon is called antiplasticization) [87–89]). However, in other cases (i.e., for the plasticizers of different type) the macroscopic properties of final composites exhibit a natural decrease characteristic of plasticization.

This is confirmed by data presented in Fig. 4.12, showing the plots of ultimate strength on compression σ_{comp} and the dynamic modulus of elasticity E' versus the concentration of various plasticizers in composites based on the epoxy oligomer ED-20, cured by diaminodiphenylmethane (DPM) and diethylenetriamine (DETA).

We will not dwell here on the mechanisms and character of plasticization in low-cross-linked polymers, which are considered in detail in the well-known monograph of Kozlov and Papkov [88].

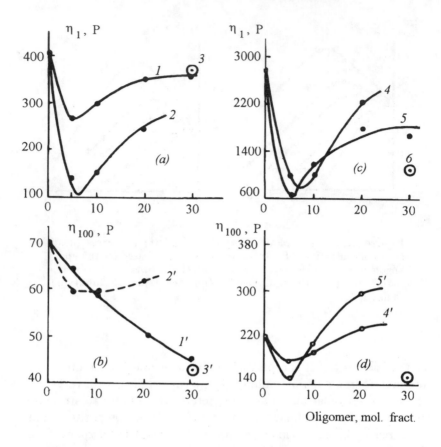

Figure 4.13. The plots of η_1 (*a, c*) and η_{100} (*b, d*) versus concentration of (*1, 3, 4, 6*) trioxyethylenedimethacrylate and (*2, 5*) an oligomer compound D-20/59: (*1–3, 1'–3'*) freshly-prepared PVC pastisols; (*4–6, 4'–6'*) after a 10-day storage; (*3,3',6,6'*) in the absence of tertbutylperbenzoyl; (*1,2,4,5*) in the presence of 1% tertbutylperbenzoyl.

The above examples, as well as all other experimental data on the plasticization of rigid and elastic oligomer-blend composites (see, e.g., [90–94]), do not contradict the conclusions following from the scheme of oligomer blend cure considered in Section 3.8. This agreement justifies using this scheme to predict variations in the properties of plasticized compositions. For example, the phenomenon of microsyneresis, involved in the cross-linking, drives nonreactive additives to the periphery of globules and into the interglobular space. The change in the viscosity of medium as a result of plasticization leads to variation in the parameters (dimensions and lifetimes) of cybotaxises. The low-molecular-mass additives in heterogeneous (e.g., bitumen–oligomer) mixtures lead to a shift in the fraction and phase equilibrium, etc. All the thermodynamic, kinetic, and structural consequences of plasticization seem to be quite reasonable when analyzed within the framework of the model of controlled structure formation in cured oligomer blends (see Section 4.1).

We have considered above the situation where a nonreactive plasticizer is used to modify a reactive oligomer blend. An apparently similar, albeit somewhat different, situation takes place on adding reactive oligomers into strongly plasticized polymers, in particular, plastisols (physical gels). Modification of PVC plastisols by reactive oligomers of various types attracts the attention of researchers for a long time and has important practical applications. In this case, technologists have to solve a number of special problems in addition to the usual ones typical of the oligomer blend processing. For example, it is necessary to know how the plastisol viscosity will change in the presence of oligomers, how the reactive components affect the properties of plastisols, what changes are to be expected in the character of monolith formation, etc. The answers can be found in numerous publications (see, e.g., [95–100]).

Below we give some examples to illustrate the effect of oligoester acrylates on the shelf life of PVC plastisols. Figure 4.13 shows a concentration dependence of viscosity of a freshly prepared PVC pastisol (with dioctylphthalate partly replaced by oligoester acrylate) and of that after a 10-day storage [99].* A difference in the character of viscosity variations measured at a shear rate of $\gamma = 1$ s^{-1} (η_1) and 100 s^{-1} (η_{100}) in the as-prepared plasisols is directly caused by a thixotropic character of the dispersion medium. The high shear rate leads to the destruction of network and the drop in viscosity. A minimum in the η_1 vs. c curves reflects the interplay between a decrease in viscosity caused by increasing content of the low-viscosity component and an increase in viscosity due to a growth in the degree of association induced by the same component.

A change in the character of the η_1 vs. c dependence observed after "aging" shows that the monolith formation process has a kinetic

---

* The total content of the liquid phase (dioctylphthalate plus oligoester acrylate) in these experiments was constant (67 wt. fract. per 100 wt. fract. of PVC emulsion).

nature and involves the reactive component. This conclusion is also confirmed by the fact that viscosity of the as-prepared plastisols is not affected by the presence of polyreaction initiators, whereas viscosity of the "aged" plastisols is markedly higher in samples charged with peroxide. Nevertheless, the complete solubility of these systems in cyclohexane indicates that chemical conversions in the reactive component during the room-temperature storage are restricted (even in the presence of initiator) to the β-polymerization. Only increasing the temperature above 130°C leads to accelerated monolith formation, which is accompanied by the chemical cross-linking and leads to better elastic–deformation properties of cured composites.

REFERENCES

1. Mezhikovskii, S.M., *Polimer–Oligomernye Kompozity* (Polymer–Oligomer Composites) Moscow: Znanie, 1989 (in Russian).
2. Mezhikovskii, S.M., Vasil'chenko, E.I., and Khotimskii, M.N., *Kompoz. Polim. Mater.*, 1987, no. 32, p. 3.
3. Mezhikovskii, S.M. and Khotimskii, M.N., *Kauch. Rezina*, 1991, no. 3, p. 10.
4. Mezhikovskii, S.M., Zhil'tsova, L.A., and Chalykh, A.E., *Vysokomol. Soedin., Ser. B*, 1986, vol. 28, no. 1, p. 42.
5. Nesterov, A.E. and Lipatov, Yu.S., *Fazovoe Sostoyanie Rastvorov i Smesei Polimerov. Spravochnik* (Phase State of Polymer Solutions and Blends. A Handbook), Kiev: Naukova Dumka, 1987 (in Russian).
6. Lipatov, Yu.S., *Fizicheskaya Khimiya Napolnennykh Polimerov* (Physical Chemistry of Filled Polymers), Moscow: Khimiya, 1977 (in Russian).
7. Kausch, H., *Polymer Fracture*, Heidelberg: Springer, 1978.
8. Kuleznev, V.N., *Smesi Polimerov* (Polymer Blends), Moscow: Khimiya, 1980 (in Russian).
9. Nielsen, L., *Mechanical Properties of Polymers and Polymeric Composites*, Moscow: Khimiya, 1978 (Russian translation).
10. Ogibalov, P.M., Lomakin, V.A., and Kishkin, B.P., *Mekhanika Polimerov* (Mechanics of Polymers), Moscow: Mosk. Gos. Univ., 1975 (in Russian).
11. Malinskii, Yu.M., in: *Kompozitsionnye Polimernye Materialy*, Kiev: Naukova Dumka, 1975, p. 83 (in Russian).
12. Bucknall, C.B., *Toughened Plastics*, London: Pergamon, 1977.
13. Rebinder, P.A., *Fiziko-Khimicheskaya Mekhanika Dispersnykh Struktur* (Physico-Chemical Mechanics of Disperse Structures), Moscow: Nauka, 1966 (in Russian).
14. Berlin, A.A., Vol'fson, S.A., Oshmyan, V.G., and Enikolopov, N.S., *Printsipy Sozdaniya Kompozitsionnykh Polimernykh Materialov* (Principles of Creation of New Composite Polymer Materials), Moscow: Khimiya, 1990 (in Russian).

15. Dickie, R.A., in: *Polymer Blends,* Paul, D.R. and Newman, S., Eds., New York: Academic, 1978.
16. Bucknall, C.B., in: *Polymer Blends,* Paul, D.R. and Newman, S., Eds., New York: Academic, 1978.
17. Lukomskaya, A.I. and Evstratov, V.F., *Osnovy Prognozirovaniya Mekhanicheskogo Povedeniya Kauhukov i Rezin* (Principles of Prediction of Mechanical Properties of Rubbers), Moscow: Khimiya, 1975 (in Russian).
18. Morton, M., in: *Multicomponent Polymeric Systems,* Moscow: Khimiya, 1974, p. 97. (Russian translation).
19. Babaevskii and Kulik, S.G., *Treshchinnostoikost' Otverzhdennykh Polimernykh Kompozitsii* (Cracking Resistance of Cured Polymer Compositions), Moscow: Khimiya, 1991 (in Russian).
20. Mezhikovskii, S.M., Structure and Properties of Polymer–Oligomer Systems and Related Composites, *Doctoral (Tech. Sci.) Dissertation,* Moscow, 1983 (in Russian).
21. Mal'chevskaya, T.D., Formation and Properties of Vulcanizates Based on Rubber–Oligomer Compositions, *Cand. Sci. (Chem.) Dissertation,* Moscow, 1980 (in Russian).
22. Frenkel', R.Sh. and Panchenko, V.I., *Modifikatsiya Rezin Oligoefirakrilatami* (Modification of Elastomers by Oligoesteracrylates), Moscow: TsNIITENeftekhim, 1981 (in Russian).
23. Rebinder, P.A., *Izbrannye Trudy. Poverkhnostnye Yavleniya v Dispersnykh Sistemakh* (Selected Works. Surface Phenomena in Disperse Systems), Moscow: Nauka, 1979 (in Russian).
24. Abramzon, A.A., *Poverkhnostno-Aktivnye Veshchestva* (Surfactants), Leningrad: Khimiya, 1976 (in Russian).
25. Pugachevich, P.P., Beglyarov, E.M., and Lavygin, I.A., *Poverkhnostnye Yavleniya v Polimerakh* (Surface Phenomena in Polymers), Moscow: Khimiya, 1982 (in Russian).
26. Veselovskii, R.A., in: *Fizikokhimiya Mnogokomponentnykh Polimernykh Sistem* (Physical Chemistry of Multicomponent Polymeric Systems), Kiev: Naukova Dumka, 1986, vol. 1, p. 250 (in Russian).
27. Sukhareva, L.A., *Poliefirnye Pokrytiya* (Polyester Coatings), Moscow: Khimiya, 1987 (in Russian).
28. Tolstaya, S.N. and Shabanova, S.A., *Primenenie Poverkhnostno-Aktivnykh Veshchestv v Lakokrasochnoi Promyshlennosti* (Application of Surfactants in Lacquer-Paint Industry), Moscow: Khimiya, 1976 (in Russian).
29. Mezhikovski, S.M. *et. al., Plaste Kautsch.,* 1979, no. 5, p. 257.
30. Berlin, A.A., in: *Dokl. 1 Vses. Konf. po Khimii i Fiziko-Khimii Polimerizatsionnosposobnykh Oligomerov* (Papers presented at the 1st All-Union Conf. on the Chemistry and Physical Chemistry of Polymerizable Oligomers), Chernogolovka: Akad. Nauk SSSR, 1977, vol. 1, p. 8 (in Russian).
31. Frenkel, R.Sh. and Mezhikovski, S.M., *Plaste Kautsch.,* 1979, no. 5, p. 257.
32. Kuleznev, V.N., *Kolloidn. Zh.,* 1976, vol. 38, no. 1, p. 175.
33. Irzhak, V.I., Rozenberg, B.A., and Enikolopov, N.S., *Setchatye Polimery* (Criss-Linked Polymers), Moscow: Nauka, 1979 (in Russian).

34. Klempner, D., Frisch, H.L., and Frisch, K.C., *J. Polym. Sci.*, *A-2*, 1970, no. 8, p. 921.
35. Sergeeva, L.M. and Lipatov, Yu.S., in: *Fizikokhimiya Mnogokomponentnykh Polimernykh Sistem* (Physical Chemistry of Multicomponent Polymeric Systems), Kiev: Naukova Dumka, 1986, vol. 2, p. 137 (in Russian).
36. Rossovitskii, V.F.and Lipatov, Yu.S., in: *Fizikokhimiya Mnogokomponentnykh Polimernykh Sistem* (Physical Chemistry of Multicomponent Polymeric Systems), Kiev: Naukova Dumka, 1986, vol. 2, p. 229 (in Russian).
37. Entelis, S.G., Evreinov, V.V., and Kuzaev, A.I., *Reaktsionnosposobnye Oligomery* (Reactive Oligomers), Moscow: Khimiya, 1976 (in Russian).
38. Entelis, S.G., in: *Dokl. 1 Vses. Konf. po Khimii i Fiziko-Khimii Polimerizatsionnosposobnykh Oligomerov* (Papers presented at the 1st All-Union Conf. on the Chemistry and Physical Chemistry of Polymerizable Oligomers), Chernogolovka: Akad. Nauk SSSR, 1977, vol. 1, p. 179 (in Russian).
39. Zuev, Yu.S., *Razrushenie Elastomerov v Usloviyakh Kharakternykh dlya Ekspluatatsii* (Degradation of Elastomers under Typical Working Conditions), Moscow: Khimiya, 1976 (in Russian).
40. Flory, P., *Principles of Polymer Chemistry*, New York: Wiley, 1953.
41. Lipatova, T.E., *Kataliticheskaya Polimerizatsiya Oligomerov i Formirovanie Polimernykh Setok* (Catalytic Polymerization of Oligomers and the Formation of Polymeric Networks), Kiev: Naukova Dumka, 1974, (in Russian).
42. Malkin, A.Ya. and Kulichikhin, S.G., *Reologiya v Protsessakh Obrazovaniyu i Prevrashcheniya Polimerov* (Rheology of the Processes of Formation and Transformation of Polymers), Moscow: Khimiya, 1985 (in Russian).
43. Malkin, A.Ya. and Begishev, V.P., *Khimicheskoe Formovanie Polimerov* (Chemical Forming of Polymers)), Moscow: Khimiya, 1991 (in Russian).
44. Lyubartovich, S.A., Morozov, Yu.L., and Tret'yakov, O.B., *Reaktsionnoe Formovanie Polimerov* (Reactive Forming of Polymers)), Moscow: Khimiya, 1990 (in Russian).
45. Leonov, L.I., Basov, N.I., and Kazankov, Yu.V., *Osnovy Pererabotki Reaktoplastov i Rezin Metodom Lit'ya pod Davleniem* (Principles of Processing Reactoplasts and Rubbers by Pressure Die Casting), Moscow: Khimiya, 1977 (in Russian).
46. Lipatov, Yu.S., *Fizikokhimicheskie Osnovy Napolneniya Polimerov* (Physico-Chemical Principles of Polymer Filling), Moscow: Khimiya, 1991 (in Russian).
47. Solomko, V.P., *Napolnennye Kristallizuyushchiesya Polimery* (Filled Crystallizable Polymers), Kiev: Naukova Dumka, 1980 (in Russian).
48. Todosiichuk, G.T. and Lipatov, Yu.S., in: *Fizikokhimiya Mnogokomponentnykh Polimernykh Sistem* (Physical Chemistry of Multicomponent Polymeric Systems), Kiev: Naukova Dumka, 1986, vol. 1, p. 130 (in Russian).

49. Lipatov, Yu.S. *et. al.*, *Vysokomol. Soedin.*, *Ser. A*, 1988, vol. 30, no. 2, p. 443.
50. Semenovich, G.M. and Lipatov, Yu.S, in: *Fizikokhimiya Mnogokompo-nentnykh Polimernykh Sistem* (Physical Chemistry of Multicomponent Polymeric Systems), Kiev: Naukova Dumka, 1986, vol. 1, p. 186 (in Russian).
51. Semenovich, G.M. *et. al.* *Vysokomol. Soedin.*, *Ser. A*, 1978, vol. 20, no. 10, p. 2375.
52. Tovmasyan, Yu.M. *et. al., Dokl. Akad. Nauk SSSR*, 1978, vol. 238, no. 4, p. 893.
53. Vostroknutov, E.T. *et. al., Pererabotka Kauchukov i Rezinovykh Sme-sei* (Processing of Rubber and Elastomer Mixtures), Moscow: Khi-miya, 1980 (in Russian).
54. Sato, Y. and Fuzukawa, J., *Rubber Chem. Technol.*, 1963, vol 36, p. 1081.
55. Knunyants, N.N., *Mekh. Kompoz. Mater.*, 1986, no. 2, p. 231.
56. Khotimskii, M.N., Mezhikovskii, S.M., and Enikolopov, N.S., in: *Per-spektivnye Elastomery i Elastomernye Kompozitsii dlya RTI* (Promis-ing Elastomers and Elastomer Composites for Technical Rubber Arti-cles), Moscow: TsNIITENeftekhim, 1981 (in Russian).
57. Sedov, L.N. and Mikhailova, Z.V., *Nenasyshchennye Poliefiry* (Unsat-urated Polyesters), Moscow: Khimiya, 1977 (in Russian).
58. Berlin, A.A., Korolev, G.V., Kefeli, T.Ya., and Sivergin, Yu.M., *Akri-lovye Oligomery i Materialy na Ikh Osnove* (Acrylic Oligomers and Related Materials), Moscow: Khimiya, 1983 (in Russian).
59. Shapoval, G.S. and Lipatova, T.E., *Elektrokhimicheskoa Initsiirovanie Polimerizatsii* (Electrochemical Initiation of Polymerization), Kiev: Naukova Dumka, 1977 (in Russian).
60. Makhlis, F.A., *Radiatsionnaya Khimiya Polimerov* (Radiation Chem-istry of Polymers), Moscow: Atomizdat, 1976.
61. Nizel'skii, Yu.N., *Kataliticheskie Svoistva Beta-diketonatov Metallov* (Catalytic Properties of Metal Beta-Diketonates), Kiev: Naukova Dumka, 1977 (in Russian).
62. Mogilevich, M.M., Turov, B.s., Morozov, Yu.L., and Ustavshchikov, B.F., *Zhidkie Uglevodorodnye Kauchuki* (Liqud Hydrocarbon Rub-bers), Moscow: Khimiya, 1983 (in Russian).
63. Mogilevich, M.M., *Okislitel'naya Polimerizatsiya v Protsessakh Plen-koobrazovaniya* (Oxidative Polymerization in the Film Formation Pro-cesses), Leningrad: Khimiya, 1977 (in Russian).
64. Maslyuk, A.F. and Khranovskii, V.A., *Fotokhimiya Polimerizatsion-nosposobnykh Oligomerov* (Photochemistry of Polymerizable Oligo-mers), Kiev: Naukova Dumka, 1989 (in Russian); Maslyuk, A.F., *Fo-topolimerizatsiya Oligomerov i Kompozitsii na Ikh Osnove* (Photopoly-merization of Oligomers and Related Compositions), Chernogolovka: Akad. Nauk SSSR, 1990 (in Russian).
65. Lazarenko, E.T., Photopolymeric Printed Forms Made of Oligoester-acrylates, *Doctoral (Tech. Sci.) Dissertation*, Moscow, 1983 (in Rus-sian).
66. Ivanchev, S.S., *Radikal'naya Polimerizatsiya* (Radical Polymeriza-tion), Leningrad: Khimiya, 1977 (in Russian).

67. Kabanov, V.A. and Topchiev, D.A., *Polimerizatsiya Ioniziruyushch-ikhsya Monomerov* (Polymerization of Ionizable Monomers), Moscow: Khimiya, 1975 (in Russian).
68. Shiryaeva, G.V. and Chikin, Yu.A., *Radiatsionnokhimicheskaya Tekh-nologiya Otverzhdeniya Kompozitsii na Osnove Oligomernykh Sistem* (Radiation-Chemical Curing Technology for Compositions Based on Oligomer Systems), Moscow: NIITEKhim, 1984 (in Russian).
69. Voronov, S.A., Tokarev, V.S., Lastukhin, Yu.A., *Geterofunktsional'nye Oligoperoksidy* (Heterofunctional Oligoperoxides), Chernogolovka: Akad. Nauk SSSR, 1990 (in Russian).
70. Rozenberg, B.A., *Problemy Fazoobrazovaniya v Oligomer–Oligomer-nykh Sistemakh* (Problems of Phase Formation in Oligomer–Oligomer Systems), Chernogolovka: Akad. Nauk SSSR, 1986 (in Russian).
71. Gurov, A.A., Mikirov, G.S., and Zhezlova, S.A., *Anaerobnye Oligo-mernye Produkty "Locktight"* (Locktight Oligomeric Anaerobic Prod-ucts), Moscow: GONTI, 1971 (in Russian).
72. Tvorogov, N.Ch., *Vysokomol. Soedin., Ser. A*, 1978, vol. 18, no. 7, p. 1461; 1983, vol. 25, no. 2, p. 248.
73. Matveeva, I.A., Zhurakovskaya, I.I., and Tvorogov, N.Ch., *Vysokomol. Soedin., Ser. A*, 1991, vol. 33, no. 10, p. 2225,
74. Bagdasar'yan Kh.S., *Teoriya Radikal'noi Polimerizatsii* (Theory of Radical Polymerization), Moscow: Nauka, 1966 (in Russian).
75. *Sostavy Anaerobnye Uplotnyayushchie (Germetiki), Klei Akrilovye, Katalog* (Anaerobic Sealing Compounds (Hermetics) and Acrylic Glues. Catalog), Cherkassy: NIITEKhim, 1988.
76. Sineokov, A.P., *Polifunktsional'nye (Met)akrilovye Monomery. Sos-toyanie i Perspektivy Razvitiya* (Polyfunctional (Meth)acrylic Mono-mers: Present-Day State and Perspectives), Chernogolovka: Akad. Nauk SSSR, 1990 (in Russian).
77. Sineokov, A.P., Murokh, A.F., and Aranovich, D.A., *Vysokomol. Soedin., Ser. B*, 1988, vol. 30, no. 10, p. 723.
78. Emanuel', N.M. and Buchachenko, A.L., *Khimicheskaya Fizika Mole-kulyarnogo Razrusheniya i Stabilizatsiya Polimerov* (Chemical Physics of Molecular Degradation and stabilization of Polymers), Moscow: Nauka, 1988 (in Russian).
79. Tokar', M.I., Stabilized Radiation-Cured Epoxy-Acrylic Polymers with Improved Working Characteristics, *Cand. Sci. (Tech.) Dissertation*, Moscow, 1989 (in Russian).
80. Pokonova, Yu.V., *Khimiya Vysokomolekulyarnykh Soedinenii Nefti* (Chemistry of High-Molecular Petroleum Compounds), Leningrad: Len. Gos. Univ, 1980 (in Russian).
81. Gorlov, Yu.P. *et. al., Zh. Prikl. Khim.,* 1988, vol. 61, no. 3, p. 522.
82. Merkin, A.P. *et. al., Azerb. Khim. Zh.,* 1984, no. 4, p. 117.
83. Merkin, A.P., Vitel's, L.E., and Mezhikovskii, S.M., in: *Rabotosposob-nost' Kompozitsionnykh Stroitel'nykh Materialov* (Working Properties of Construction Composite Materials), Kazan': Kaz. Inzh.-Stroit. In-st., 1982, p. 35 (in Russian).
84. Nadzharyan, S.N., Bitumen-Oligomeric Compositions for the Con-struction Materials), *Cand. Sci. (Tech.) Dissertation,* Moscow, 1991 (in Russian).

85. Berlin, A.A. *et. al.*, *Khimiya Polisopryazhennykh Sistem* (Chemistry of Polyconjugated Systems), Moscow: Khimiya, 1972 (in Russian).
86. Berlin, A.A., *Usp. Khim.*, 1975, vol. 44, no. 4, p. 502.
87. Khozin, V.G., Physico-Chemical Modification of Epoxy-and Furan-Based Polymers and Development of Related Composites), *Doctoral (Tech. Sci.) Dissertation,* Leningrad, 1980 (in Russian).
88. Kozlov, P.V. and Papkov, S.P., *Fizikokhimicheskie Osnovy Plastifikatsii Polimerov* (Physico-Chemical Principles of Polymer Plasticization), Moscow: Khimiya, 1982 (in Russian).
89. Perepechko, I.I., *Akusticheskie Metody Issledovaniya Polimerov* (Acoustic Methods for the Investigation of Polymers), Moscow: Khimiya, 1973 (in Russian).
90. Bol'shakov, A.I., Kiryukhin, D.P., and Barkaslov, I.M., in: *Radiatsionnaya Khimiya i Tekhnologiya Oligomernykh Sistem* (Radiation Chemistry and Technology of Oligomer Systems), Moscow: NIITEKhim, 1989, p. 31 (in Russian).
91. Garipov, N.G., Materials Based on Polymer-Oligomer Systems: PVC–Polyfurans, *Cand. Sci. (Tech.) Dissertation,* Moscow, 1987 (in Russian).
92. Kotlyar, N.A., Preparation and Properties of Epoxy-Acrylic Compositions for Low-Pressure Die Casting, *Cand. Sci. (Tech.) Dissertation,* Dnepropetrovsk, 1987 (in Russian).
93. Zadontsev, B.G., Zaitsev, Yu.S., and Yaroshevskii, S.A., *Fiziko-KhimicheskieOsnovy Materialovedeniya Polimer-Oligomernykh Kompozitov* (Physico-Chemical Principles of Materials Science of Polymer–Oligomer Composites), Chernogolovka: Akad. Nauk SSSR, 1986 (in Russian).
94. Litvinova, T.V., *Plastifikatory dlya Rezinovogo Proizvodstva* (Plasticizers for Rubber Industries), Moscow: TsNIITENeftekhim, 1981 (in Russian).
95. Gorshkov, V.S. *et. al.*, *Plast. Massy*, 1981, no. 1, p. 25.
96. Balakirskaya, V.L. *et. al.*, *Plast. Massy*, 1983, no. 4, p. 17.
97. Mezhikovski, S.M. *et. al.*, *Plaste Kautsch.*, 1978, no. 8, p. 444.
98. Muratova, L.N., Akutin, M.S., and Il'in, S.N., *Plast. Massy*, 1983, no. 10, p. 11.
99. Gorshkov, V.S. *et. al.*, *Vysokomol. Soedin., Ser. A*, 1978, vol. 20, no. 2, p. 402; *Ibid*, 1978, vol. 20, no. 6, p. 1369; *Ibid*, 1979, vol. 21, no. 5, p. 1091.
100. Yaroshevskii, S.A., Polymer–Oligomer Composite Materials Based on Linear Polymers and Oligo(ester acrylates). Features of the Preparation and Processing Technology, *Cand. Sci. (Tech.) Dissertation,* Moscow: Inst. Khim. Fiz., Akad. Nauk SSSR, 1986 (in Russian).
101. Lomonosova, N.V., *Vysokomol. Soedin., Ser. A*, 1992, vol. 34, no. 6, p. 49.
102. Mezhikovskii, S.M., in: *Papers Presented at the Intern. Rubber Conf. (IRC-94),* Moscow: 1994, vol. 3, p. 83.
103. Ionov, Yu.A., Morozov, Yu.L., and Reznichenko, S.V., in: *Papers Presented at the Intern. Rubber Conf. (IRC-94),* Moscow: 1994, vol. 3, p. 61.

CONCLUSION

There are numerous advantages of using oligomers in the synthesis of polymeric materials. The efficiency of commercial processes increases because of decreasing power consumption and the possibility to use molding technologies. The quality of final materials improves as a result of chemical and physical modification of the polymer matrix. The costs of the products can be reduced by introducing excess mineral fillers without danger of decrease in the working properties. Excluding stages involving volatile organic solvents increases the ecological safety of the technological process, etc.

However, the current state-of-the-art in the physical chemistry of oligomeric systems is still far from the level that would allow us quantitatively assess and reliably predict all correlations between the composition of the initial components (raw materials), the structure of a liquid phase (intermediate products), and a combination of the working properties of cured material (article). The physicochemical laws governing the conversion of liquid oligomer blend systems into solid polymeric materials, considered in this monograph, are only outlining the boundaries of the information space in which a technologist has a freedom to "perfect" the final material to reach a necessary condition (depending on the purpose) without a risk of unjustified labor and material consumption. Indeed, the knowledge of the physicochemical laws controlling the technological process markedly decreases the number of experiments necessary to optimize the material properties and increases the possibility to employ all the useful properties inherent in the oligomer nature that were previously lost on the pathway from a raw oligomeric material to final polymer.

For example, it was a commonly accepted notion until recently (and is still not a seldom belief now) that the best physicochemical properties of polymer blends are achieved by homogeneously dispersing the components, which explained the efforts made to ensure the obtaining of ultimately homogeneous mixtures. Now it is obvious that there are alternative, more efficient ways to increase the elastic-deformation properties of polymeric articles, in particular, by ensuring a definite level of heterogeneity. Above we have demonstrated how, once the thermodynamic parameters of the blend are known, to control by simple technological methods the morphology of composites formed upon the cure of oligomer blends.

Another example is the permanent wish of technologists to decrease the viscosity of compositions processed, which is a quite natural trend in view of reduced energy consumption. Thus, when the phenomenon of "anomalous" increase of the polymer viscosity in the presence of oligomeric additives was discovered, the first reaction of technologists was to exclude the effect by selecting the blend compositions and temperature regimes so as to avoid the viscosity growth manifestations. However, this was a logical but trivial course of action. A thorough physicochemical analysis of the problem showed that this, apparently undesirable, effect can be used to increase the material quality. For example, by introducing proper amounts of oligomeric additives (thus setting a necessary viscosity level ensuring uniform distribution of mineral fillers over the matrix volume), we may considerably improve some physicomechanical properties of the filled composites. The energy losses, connected with increased viscosity, are compensated in the subsequent processing stages by adding reactive plasticizers. This technological method was also described above.

It is possible to proceed with numerous examples of similar, apparently paradoxical, applications of the physicochemical effects related to oligomer blends. However, we should prefer to emphasize one feature in common for all oligomeric blend systems, which has been yet not paid proper attention. This concerns a problem of priority between thermodynamic and kinetic factors in the practical technology. Until recently, if it was necessary to change some macroscopic property of a system, it was a usual course of action for technologists to control one of the "thermodynamic" variables (temperature, pressure, etc.), that is, to change the parameters of state so that the system would "fall" within a definite zone of the phase diagram. This is a quite justified approach, since the phase transition is usually accompanied by a jumplike change in properties of the system. However, the problem is that oligomeric systems have characteristic times of the relaxation to equilibrium that are comparable with a typical duration of some stages involved in the technological process. Thus, we must know whether the structural transformation of the oligomer blend will be completed (to reach a thermodynamically allowed level ensuring the desired properties) by the moment of, e.g., switching on the mechanism of cure. Unfortunately, this problem was not paid a proper attention in practice, although an allowance for (or neglect of) the kinetic factors may not only change the quantitative characteristic (which is certainly of value), but modify the qualitative character

W_0, min^{-1}

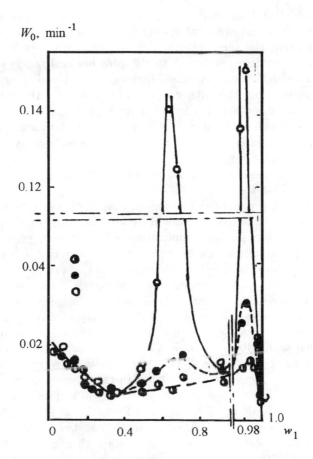

The plot of W_0 versus the oligomer content w_1 for a PVC–trioxy-ethylenedimethacrylate blend for $\tau_{exp} = 1$ (*1*); 5 (*2*); 40 days (*3*).

of the observed trends as well (which can be of much greater impor-tance).

This is clearly illustrated in the figure, which was specially saved for the final discussion. The figure shows typical curves of the initial rate of cure of an oligomer blend versus the concentration, measured at various storage times of a PVC–oligoester acrylate blend. As is seen, the jumplike increase in W_0 in the metastable region and in the vicinity of the critical point is pronounced only for $\tau_{exp} \rightarrow \tau_{exp}^{cr}$, otherwise the effects are insignificant or virtually absent. It is an independent problem that the τ_{exp}^{cr} values inherent in this system are too large from the practical standpoint: the value can be decreased by increasing the temperature of exposure. But this is already a particular problem pertaining to the possible realization of general physicochemical laws established for oligomeric blend systems.

In conclusion, the author would like to attract the attention of readers to some considerations of more general character, related to the physicochemical features of oligomer blends. Although these problems are apparently not directly pertaining to technology, but nevertheless...

We will speak of the possible role of oligomeric systems in the origin of life on the Earth.

If we will proceed from an axiom that the life has spontaneously originated by evolution of carbon from atomic to high-molecular state, and developed by further complication to biopolymer systems capable of self-replication (in case of supernatural pathways, it is still of interest as to how the Creator did manage this). In order to rationalize this extremely involved process, it is necessary to answer numerous questions to which the efforts of generations of scientists were devoted for many centuries. We will naturally not try to discuss all of these questions, which are subjects of numerous special publications, but will touch upon a single problem as to why the evolution proceeding in a general chaos of the World Ocean shifted toward the formation of a complex organic matter, although the laws of thermodynamics suggest that any increase in the complexity of a close system is unfavorable. In 1988 R.Fox ("Energy and Evolution of Life on the Earth", San Francisco: Freeman, 1988) remarked that it is "unclear how could large molecules appear on the Earth, since they have a tendency to decay as rapidly as form. This gives rise to an Uroborus type puzzle: how could the polymer assembly start if this were possible only in the presence of polymers."*

In 1987, academician V. Gol'danskii pointed out four stages in the origination of life: (1) formation and accumulation of monomers; (2) differentiation of monomers with respect to chiral purity; (3) formation of oligomers; (4) synthesis of biologically active macromolecules. It was assumed that the latter are formed of oligomers, and the energy barrier is surmounted (according to R. Fox) using the energy of oligomer activation to the state in which they become capable of spontaneous polymerization. This is a possible course of events, but how and why?

---

* Uroborus is a mythical dragon swallowing its own tail, thus symbolizing structures having neither origin no end.

In this monograph, the attention of readers was attracted to some apparently anomalous effects, manifested by oligomer blends under certain conditions. In particular, two important circumstances are especially worth of noting:

(i) Oligomeric systems have a tendency to self-organization with the formation of associates and cybotaxises, which is not inspired by an external driving force, but is rather a result of minimization of the system energy through weak intermolecular interactions between the oligomer molecules proper;

(ii) If the parameters of state (temperature, concentration, pressure) acquire the values corresponding to a critical point or a metastable region, the system exhibits a sudden increase in the structural order, accompanied by a sharp jump in the viscosity and the initial polymerization rate.

Taking into account the postulates of V. Gol'danskii and R. Fox, we may admit that oligomers appear in some part of the Universe at a definite stage of its evolution (which is essentially a stochastic process!). The oligomer molecules exhibit self- organization to form associates, cybotaxises, or mixed (according to R. Fox) oligomers. These formations are characterized by a sufficiently high probability of catalytic, radiation-stimulated, or some other process switching on the initiation or activation mechanism (polymerization, polycondensation, or polyaddition) leading to the formation of polymer molecules. This process naturally involves the reverse reactions (depolymerization, dehydration, degradation, etc.), accompanied by the decay of polymer molecules. Thus, we obtain a dynamic system:

$$\text{Monomer} \Longleftrightarrow \text{Oligomer} \Longleftrightarrow \text{Polymer},$$

in which the equilibrium may well shift to the left.

We can readily imagine that the external conditions change at some moment of time, and the equilibrium system will occur in a situation where the temperature, pressure, and concentration correspond to a critical or metastable state. During a many-century history of the Earth, a much longer evolution of Galaxy, and in the eternal Universe, the situation when all the above conditions were realized could definitely take place sometime and somewhere. Once this happened, the system must unavoidably exhibit an anomalous increase in the viscosity. This, in turn, would lead to an anomalous growth in the initial polymerization rate and the accumulation of the polymer would give rise to the well-known gel effect. Under these conditions, the chemical process of the formation of macromolecules must proceed at an increasing rate and become irreversible. Of course, it was only a part of the macromolecules, formed in this way, that could eventually lead (for various reasons) to the formation of biologically active polymers. This is a different question, related to fine mechanisms of the synthesis and subsequent evolution of both individual macromolecules (of various origin) and the system as a whole.

The idea that anomalies of the physicochemical behavior of the monomer–oligomer–polymer compositions may probably account for "violation" of the thermodynamic laws in the blend systems, as a model of origination and development of the high-molecular substance in the World Ocean, still requires proper justification and corroboration. Nevertheless, even in the schematic form outlined above, these notions seem to be quite reasonable. Indeed, the idea is based on a real material, does not contradict concepts put forward by the other natural sciences, and is logical and attractive. The latter is an important evidence, since everything in both nature and science must be beautiful and harmonic.